NATIONAL BIRDS OF THE WORLD

Ron Toft

B L O O M S B U R Y
LONDON • NEW DELHI • NEW YORK • SYDNEY

Dedicated to the memory of Jean Toft,
the author's late wife,
who died during the writing of this book.

First published in 2014

Bloomsbury Publishing Plc, 50 Bedford Square, London WC1B 3DP

www.bloomsbury.com

Bloomsbury Publishing, London, New Delhi, New York and Sydney

A CIP catalogue record for this book is available from the British Library

Commissioning editor: Julie Bailey
Project editor: Jasmine Parker
Design: Nicola Liddiard, Nimbus Design

ISBN 978-1-4081-7835-5

Printed in China by C&C Offset Printing Co Ltd.

10 9 8 7 6 5 4 3 2 1

NATIONAL BIRDS OF THE WORLD

Ron Toft

BLOOMSBURY
LONDON • NEW DELHI • NEW YORK • SYDNEY

Contents

Foreword

For me birds are a way of easily and accessibly engaging with a diversity of life on a daily basis. Thus my favourite birds are those that live around me, in my back garden.

Of course such an interest in birds means that I also have an interest in seeing new or exotic species, and unravelling their individual behaviours and ecologies, as well as their simple beauty, enhances my quality of life. And that is no bad thing! Nor unique.

Wherever I travel in the world I meet people to whom birds mean something, and this is nothing new either. For centuries we have observed, studied, worshipped and revered birds. We have chosen them as icons and symbols with human parallels and very often allowed our anthropomorphising to run riot, producing some unlikely and unsuitable heroes and generating all sorts of misconceptions.

This book explores birds very much as symbols, and they have been selected over varying degrees of time and with varying degrees of care and biological accuracy. Nevertheless these avian totems mean something to those whose ideas and aspirations they represent, and here we see how and why such choices were made.

I must confess I wouldn't mind a rethink in some cases. I think it would be a good idea if nations, states and regions now selected a bird in some kind of conservation crisis. This would draw attention to its plight and perhaps also help fuel its protection. This certainly helped the North American Bald Eagle when its numbers had become precarious. Maybe it would do the same for others. So forget the Robin. How about having the Lapwing as the British national bird!

Chris Packham

Bermuda Petrel, IUCN: **EN**

Giant Ibis, IUCN: **CR**

Grenada Dove, IUCN: **CR**

Javan Hawk-eagle, IUCN: **EN**

Saker Falcon, IUCN: **EN**

Montserrat Oriole, IUCN: **CR**

Kagu, IUCN: **EN**

Philippine Eagle, IUCN: **CR**

Tooth-billed Pigeon, IUCN: **EN**

St Helena Plover, IUCN: **CR**

Imperial Amazon, IUCN: **EN**

Red-crowned Crane, IUCN: **EN**

IUCN – International Union for Conservation of Nature.

The IUCN Red List of threatened species is a classification system, which the IUCN launched to raise the profile of threatened animals on our planet. There are nine categories: Not evaluated (NE); Data deficient (DD); Least concern (LC); Near threatened (NT); Vulnerable (Vu); Endangered (EN); Critically endangered (CR); Extinct in the wild (EW) and Extinct (EX). Throughout the book each national bird has its IUCN classification listed.

Introduction

When I began conducting research for this book in early 2012, I naively thought that ascertaining which countries/territories had a national bird would be a relatively straightforward exercise. I remember thinking, too, that if a country did have an avian icon, surely someone in authority would know when and why a particular species was adopted, and that this information would be documented somewhere and readily available.

With a few notable exceptions, however, nothing has been easy or straightforward. Many Internet references for national birds are partially or completely inaccurate, emails to embassies and government departments and agencies have often gone unanswered, despite several attempts at establishing contact, and the information I have managed to unearth has sometimes been frustratingly inadequate. Perseverance, though, has paid off. I owe a considerable debt of gratitude to all the conservationists around the world who came to my rescue. If they were not able to help me directly in my quest, they usually knew a man, or woman, in their country who could.

Given the paucity of information about national birds in many countries, it may well be that some have inadvertently been overlooked. If that proves to be the case, I apologise now and ask the countries in question to contact the publisher so that these omissions can be rectified in future editions of this book. Of the world's 200 or so nations, half have a national bird. Many avian icons, however, have never been officially chosen or endorsed by a government. Some birds have been adopted by conservation bodies; others as a result of public-consultation exercises.

There are some well-known countries, notably Canada, Ireland and the Netherlands, which do not have a national bird for some reason, and relatively little-known island states, such as Samoa and Palau, which most definitely do.

The oldest national bird is the USA's Bald Eagle, which was adopted by Congress as long ago as 1782 despite attempts by Benjamin Franklin to confer this status on the Wild Turkey. Franklin considered the Bald Eagle to be a bird of 'bad moral character' and the Wild Turkey to be 'a much more respectable' species.

The newest national birds appear to be Scotland's Golden Eagle, voted for in 2013; Mongolia's Saker Falcon, adopted in 2012; and Yemen's Arabian Golden-winged Grosbeak and Israel's Eurasian Hoopoe, both adopted in 2008. At least two countries/territories seem set to choose a national bird in the near future. Angola is likely to adopt the Red-crested Turaco, while Palestine favours the Palestine Sunbird.

Trinidad and Tobago is the only nation with not one but two completely separate official national birds – the Scarlet Ibis for Trinidad and the Rufous-vented Chachalaca for Tobago, while the Cayman Islands' national bird is two subspecies of the Cayman Parrot. Mauritius is unique in having adopted an extinct species, the celebrated Dodo, as its avian icon.

The most popular species are clearly the Andean Condor, which has been adopted by Bolivia,

Colombia, Chile and Ecuador, and the African Fish-eagle, which is the avian icon of Malawi, Namibia, Zambia and Zimbabwe.

Given the wealth of exotic, colourful species in Argentina, Brazil and Costa Rica, I found it surprising that all three countries have adopted fairly drab birds – the Rufous Hornero in Argentina, the Rufous-bellied Thrush in Brazil and the Clay-coloured Thrush in Costa Rica.

Perhaps the most bizarre-looking national bird is Guyana's primeval Hoatzin or Stinkbird, an avian punk rocker, while the most beautiful birds must surely be Papua New Guinea's Raggiana Bird-of-paradise and Guatemala's Resplendent Quetzal.

Some national birds have very high profiles and are used not only officially, on items such as flags, coats of arms, coins and banknotes, but also as logos for conservation and commercial bodies. By way of contrast, some national birds seem to have very low profiles indeed, begging the question: why bother adopting them in the first place?

I have been interested in birds and have written about them for many years, and researching and writing this book has been a labour of love. I have learned many things, not least that birds are firmly embedded in the cultures of many countries and peoples.

If nothing else, I hope this book serves to illustrate the diversity and beauty of the world's birds, and how vital it is that we protect them and all the other wildlife with which we share this wonderful, precious planet.

Top: *Raggiana Bird-of-paradise*

Above: *Resplendent Quetzal*

Opposite: *Hoatzin*

Ron Toft

Winchester, January 2014

Red-crested Turaco

Tauraco erythrolophus

The Red-crested Turaco is unofficially the national bird of Angola. It has also been adopted as the emblem of a new bird conservation group in Angola and its image is used on an awareness-raising poster for this species.

WHERE TO SEE The Red-crested Turaco is quite commonly seen along the Angolan escarpment and adjacent forested areas, and has been recorded in the following Important Bird Areas (IBAs) in Angola: Calandula on the Lucala River, Camabatela, Chongoroi, Gabela and Quicama.

LEAST CONCERN

IUCN: It is not known how many mature individuals there are in the wild, but BirdLife International states that the population appears to be declining as a result of habitat destruction.

Size: Length 40–43 cm (1.4–1.5 in).

Description: Very colourful. Adults have a crimson head, nape and crest (some crest feathers are white tipped), a green back and breast, a green-black belly, a dark blue/purple tail, black legs and feet and a yellow bill.

Diet: Little is known about this species' diet, other than that it eats fruits including berries.

Reproduction: Poorly understood.

Range: Endemic to Angola.

Habitat: Moist lowland forests.

Cultural presence

Various conservationists, including ornithologist and specialist bird-tour guide Michael Mills of Birds Angola, are determined to make the Red-crested Turaco the avian icon of Angola. 'Currently there is no real national bird, so it shouldn't be too difficult!' said Mills, who has worked in Angola since 2005 on various bird projects for the A. P. Leventis Ornithological Research Institute of Nigeria in partnership with BirdLife International. 'Our strategy is to start calling the Red-crested Turaco Angola's national bird and continue to treat it as such at every opportunity. I think it will become the national bird when most people recognise it as such.'

Top: *The Red-crested Turaco appeared on this fauna and flora stamp in 1996.*

Above: *The striking Red-crested Turaco is becoming a national conservation icon in Angola where it is an endemic species.*

Zenaida Dove

Zenaida aurita

LEAST CONCERN

IUCN: The overall population is very large and estimated to comprise anything from half a million to five million mature individuals.

Size: Length 28–30.5 cm (11–12 in).

Description: A small, attractively marked dove. Adult males are a smorgasbord of cinnamon, buff-brown, rufous-brown and similar hues. They also have an iridescent purple neck patch, blackish-brown primary feathers, a black bill and red legs and feet. There are three subspecies, the one found on Anguilla being the nominate *Z. a. aurita*.

Diet: Fruits and seeds.

Reproduction: Two eggs are usually laid. Nests mainly in trees and shrubs, and on the ground on islands where there are few or no predators.

Range: Found throughout the West Indies and part of Mexico (the north coast of the Yucatan Peninsula and Cozumel Island).

Habitat: Varied, open woodland, second-growth forest, scrub and mangrove areas.

The Zenaida Dove was voted national bird in December 1993. Its defeated opponents were the Brown Pelican and the American Kestrel.

WHERE TO SEE Given its large population, the Zenaida Dove is relatively easy to spot in a wide range of environments from woodland to mangroves.

Cultural presence

When is a national bird not a national bird? When it's a 'turtle-dove' or a 'common ground-dove', as far as Anguilla is concerned. Although the Zenaida Dove is officially Anguilla's national bird, it is sometimes mistaken for and even described on certain websites as being an unspecified 'turtle-dove', a 'common ground-dove' or the similar Mourning Dove *Zenaida macroura*. According to the *Handbook of the Birds of the World*, the Zenaida Dove is intensively hunted as a game bird.

Above: *The Zenaida Dove appeared on a set of 1995 Easter stamps in Anguilla.*

Magnificent Frigatebird

Fregata manificens

The Magnificent Frigatebird is the national bird of Antigua and Barbuda. The male has a red gular sac (seen above), which can be inflated to attract a female in courtship. These birds are excellent in flight, but their enormous wingspan, short legs and small feet make walking on land appear awkward. The silhouette of a Magnificent Frigatebird is often described as prehistoric.

WHERE TO SEE The best place to see this species is the aptly named Frigate Bird Sanctuary in Barbuda's 3,600-hectare (90-acre) Codrington Lagoon National Park, where around 5,000 Magnificent Frigatebirds breed – the biggest nesting colony in the Caribbean. The birds are very approachable and are a major tourist attraction. Frigatebirds also breed on the small uninhabited island of Redonda, 46 km (29 miles) from Antigua.

LEAST CONCERN

IUCN: It is thought that globally there are probably several hundred thousand birds. The population is believed to be increasing.

Size: Length 89–114 cm (2.11–3.09 in).

Description: Prehistoric-looking but graceful seabird with a long, deeply forked tail, a large, pale, hooked bill and long, narrow, strongly 'kinked' wings. Unlikely to be confused with any other bird in flight, except perhaps for another frigatebird species. Males are largely black with an inflatable bright red throat sac, or gular pouch, while the larger females can be identified on the wing by their prominent white breast band.

Diet: Mainly flying fish and squid, but also takes other fish, jellyfish, baby turtles, seabird eggs/chicks and fish waste.

Reproduction: A single egg is laid on a stick platform, often in mangroves. Nests in colonies.

Range: Huge. One of five frigatebird species, the Magnificent Frigatebird is found from California to Ecuador on the Pacific coast of North and South America, and Florida to southern Brazil on the Atlantic coast of these continents. There is also a relict population in the Cape Verde archipelago off the west coast of Africa.

Habitat: Marine – open tropical and subtropical seas, coastlines and small islands.

Left: *The Magnificent Frigatebird is not just the biggest of the five frigatebird species but also the one with the longest bill.*

Above: *Only one frigatebird chick is raised at a time. Vulnerable to predation during the first few weeks of its life, the youngster is closely watched and protected by its parents.*

Frigatebirds and fishermen have long enjoyed a mutually beneficial relationship. The birds follow fishing boats and congregate in fishing ports ready to snap up offal and other fish scraps, and the fishermen use the birds as aerial markers as to where shoals of fish are likely to be found.

Cultural presence

Magnificent Frigatebirds are also called man-o'-war birds, perhaps because the manner in which they attack and steal from other birds drew comparisons with the actions of Caribbean pirates in bygone ages, who used frigates or man-o'-war warships to board and plunder merchant vessels.

Caribbean fishermen and hunters once harvested frigatebird eggs and young to supplement their meagre diets, according to the *Handbook of the Birds of the World*. They were also used in traditional medicine and voodoo.

The Magnificent Frigatebird has adorned many of Antigua and Barbuda's colourful stamps over the years, including five in 1994 alone. It has also appeared on a highly collectable, limited edition, 23-carat gold-foil 30-dollar banknote.

The Magnificent Frigatebird was one of several bird species featured on the logo used for the Society for the Conservation and Study of Caribbean Birds' 17th regional meeting, held in 2009 in Antigua.

This 15c Magnificent Frigatebird stamp was one of a four-value marine life set issued on Antigua & Barbuda in 1985.

ARGENTINA

Rufous Hornero

Furnarius rufus

LEAST CONCERN

IUCN: Number of mature individuals unknown, but believed to be common and increasing.

Size: Length 16–23 cm (6.3–9.05 in).

Description: Reddish-brown above and buff-brown below. Whitish throat. Sexes are alike. One of 236 species in the Furnariidae or ovenbird family.

Diet: Wide range of invertebrates, including beetles, ants, termites, grasshoppers, worms, spiders and snails.

Reproduction: Two to four eggs are laid in an adobe mud 'oven'.

Range: Found over much of South America. Native to Argentina, Bolivia, Brazil, Paraguay and Uruguay.

Habitat: Scrub, pasture and farmland as well as parks and gardens.

Adopted as Argentina's national avian icon, the Rufous Hornero is a familiar species throughout its large range and seems to have adapted well to man-made environments.

WHERE TO SEE This species is encountered in a range of habitats – from open countryside to parks and gardens in urban areas – and commonly seen near areas of human habitation. In the countryside, the Hornero likes places where there is disturbed, bare soil and is often spotted on farmland and in scrub-covered terrain.

According to the *Handbook of the Birds of the World*, the Hornero is found mostly in lowland habitats but also occupies suitable areas in mountains up to an altitude of 3,500 metres.

The Rufous Hornero is essentially a two-tone bird – shades of brown above and paler below.

Left: *The Rufous Hornero builds a clay nest typically on a tree branch or a man-made structure such as a fence or building.*

Below: *If you see a Rufous Hornero on the ground, it is likely to be bathing or walking with a fairly upright posture.*

Cultural presence

The Rufous Hornero has appeared on at least one stamp in Argentina and on the country's 25-centavos coins.

The Rufous Hornero is especially noted for its large, robust, roughly circular mud nests. Made of mud, clay, plant material and/or some dung, these so-called 'ovens' are elaborate, long-lasting structures. They are 20 to 30 cm (8–12 in) in diameter with walls that are 3 to 5 cm (1.18–2 in) thick, and weigh roughly 3 to 5 kg (6.6–11 lb). Although this species very rarely uses the same oven nest twice, it often builds a new nest on top of an old one. Its classic adobe construction method has long been used by people of various cultures around the world.

Below left: *Room with a view – this enterprising Hornero has chosen a Pantanal sign as its nest site.*

Below right: *This 1966 Argentinian stamp depicts a pair of Rufous Horneros at their mud nest.*

AUSTRALIA

Emu

Dromaius novaehollandiae

The Emu is widely regarded as Australia's official bird, despite never having been formally designated as such by the government. Both the Emu and the Red Kangaroo are depicted on Australia's coat of arms, which appears on all manner of things from passports and government insignia to the country's 50-cent coins. It is said that these two animals were chosen because they are unable to walk backwards and are thus symbolic of a young nation progressing and moving forwards.

The Emu has appeared on a variety of stamps, including a 2-pence 1888 stamp marking New South Wales' 100th anniversary, a 6-pence 1913–14 engraved definitive, a 5-pence 1930 surcharge stamp, a 1.35-dollar 1994 wildlife stamp and a 55-cent 2010 Australian commemorative coinage stamp.

WHERE TO SEE The Emu is found in most parts of Australia. It is also commercially farmed to produce low-fat meat, oil and leather.

Newly hatched chicks have distinct stripes for camouflage; they lose these after about five to six months.

LEAST CONCERN

IUCN: There are estimated to be 630,000 to 725,000 mature individuals in the wild.

Size: Length 1.5–1.9 m (5–6.2 ft); weight 30–45 kg (66.1–99.2 lb).

Description: Shaggy, flightless, large-bodied and long-legged bird. Dark brown/grey-brown in colour. The third largest bird in the world after the Ostrich and Southern Cassowary. The Emu's big, powerful, three-toed feet enable it to cover ground easily and quickly at speeds of up to 48 km (29.8 miles) per hour. This species is the only surviving member of its genus.

Diet: An omnivorous species, the Emu eats seeds, fruits, flowers and tender roots, plus lots of insects and even small invertebrates. Stones up to 46 g (1.6 oz) in weight are swallowed to help grind food in the gizzard.

Reproduction: Pairing takes place in December and January. Five to 15 large teal to dark-green-coloured eggs weighing 450–650 g (15.9–22.9 oz) are laid in a shallow, vegetation-lined depression on the ground. Incubation lasts about eight weeks. Eggs and chicks are looked after by the male.

Range: Endemic to Australia.

Habitat: Occurs in a range of habitats, especially open sclerophyll forests ('sclerophyll' is a general term for hard-leaved vegetation such as acacias and eucalyptuses) and semi-arid plains. Emus tend to stay clear of deserts and heavily populated areas, and are not found in tropical rainforests.

Cultural presence

The Emu has long been an integral part of the culture, traditions and beliefs of Aboriginal Australians. One creation myth tells of a heaven-dwelling Emu spirit who felt that the newborn Earth was too beautiful to be illuminated solely by starlight, so she tossed one of her large eggs into the sky and pierced it with her bill. The golden yoke that poured forth across the sky became Earth's first sunrise.

Some of the dust lanes in the Milky Way galaxy, on the outer edges of a spiral arm where our solar system is situated, are thought by many people to resemble an Emu. In 2005 a painting by Charmaine Green of the so-called Emu in the Sky of the Wadjiri people of Western Australia was featured on the cover of the Australian astronomy community's 10-year plan.

Cinematographer Barnaby Norris won third prize in the 2007 New Scientist Eureka Prize for Science Photography for his picture of the Emu-like galactic dust lanes rising above an ancient rock engraving of an Emu in Ku-ring-gai Chase National Park, near Sydney. The Emu in the Sky is so large that Norris's image had to be pieced together from 520 individual images. It appears upright in the night sky above the rock engraving only once a year – at the time when Emus lay their eggs.

Emu-warning road signs are a common sight in the Australian countryside, where the birds are prone to wander onto the roads.

Emu eggs, in fact, were an important food for the Ku-ring-gai and other Aboriginal Australians. The egg yolks were mixed with ochres by indigenous peoples to make paints for body art and bark painting, while the eggshells were ground up for medicinal purposes. Blown Emu eggs have been and still are turned into beautiful works of art through painting or carving.

Emu-egg carving – an art form practised by both Aboriginal and non-Aboriginal people – first became popular in the mid to late 19th century. Some silversmiths designed elaborate holders for eggs. Egg carving was kept alive in the 20th century by Aboriginal people in south-east Australia and the Carnarvon region of Western Australia.

The Australian Museum in Sydney states that there are still dozens of carvers who depict people, traditions, animals and scenes past and present on Emu eggs. It adds that these egg images provide a 'rich visual history'. The designs of the carvings vary between language groups.

Both the Australian Museum and the National Museum of Australia in Canberra have many examples of Aboriginal artefacts, including those incorporating Emu feathers. NMA exhibits include a ceremonial dance 'hat' from Mornington Island, a dance wand or 'fan' from Arnhem Land and a basket from South Australia incorporating feathers. Among non-feather artefacts are a tureen decorated with floral sprays, a kangaroo, an Emu, a rose, a thistle and a shamrock made for the Union Club of Hobart, Tasmania, and a pre-1918 gorget or breastplate featuring both an Emu and a kangaroo. Gorgets – also known as king plates – were usually awarded to 'helpful' Aboriginal people from the time of settlement onwards.

Right: *Dance hat from Mornington Island made from bark, hair, string and Emu feathers from the National Museum of Australia.*

Below: *The Emu has appeared on everything from coins and stamp to passports.*

Emu plumes were used to decorate the khaki fur felt or slouch hats worn by the elite mounted troops of the Australian Light Horse. Originally such feather-adorned hats could be worn only by Queensland's mounted troops in recognition of their service during the 1891 Great Shearers' Strike. Emu feathers still adorn the hats and berets of some Australian regular Army and reserve armoured units.

State and territory emblems

Gang-gang Cockatoo

Kookaburra

AUSTRALIAN CAPITAL TERRITORY

Gang-gang Cockatoo *Callocephalon fimbriatum* Female birds are uniformly grey.

Gang-gangs, which usually pair for life, especially like eucalypt and wattle seeds but also eat nuts, berries, fruits and insects. The call of this species has been likened to that of a squeaky gate.

NEW SOUTH WALES

Laughing Kookaburra *Dacelo gigas* The world's biggest kingfisher at around 45 cm (18 in) , the Laughing Kookaburra eats not only fish but also reptiles, rodents, worms and insects. Large prey items, such as lizards and snakes, are thrashed on a hard surface to kill and soften them before they are eaten. Its bill can grow up to 10 cm (4 in) The Laughing Kookaburra is native to eastern mainland Australia.

NORTHERN TERRITORY

Wedge-tailed Eagle *Aquila audax* This species – Australia's largest bird of prey – has been the Northern Territory's avian emblem since the region attained self-governing status in 1978 and was granted a coat of arms by Queen Elizabeth II.

Wedge-tailed Eagle

The crest features a Wedge-tailed Eagle with its wings raised, grasping a sacred Aboriginal stone or *tjurunga* on a wreath of the colours. The supporters are Red Kangaroos, one holding a Spider Conch and the other a True Heart Cockle. Beneath is a grassy, sandy mound covered with Sturt's Desert Rose flowers.

QUEENSLAND

Brolga *Grus rubicunda* Although the Brolga – one of 14 Crane species in the world – has been depicted on Queensland's coat of arms since 1977 (the year in which Queen Elizabeth II celebrated her Silver Jubilee), it wasn't declared as the state's official bird until 1986.

SOUTH AUSTRALIA

Piping Shrike or White-headed Magpie *Gymnorhina tibicen telonocua* A noted songster and very common bird in South Australia, the black and white Piping Shrike has been the state's bird emblem since 1901 when Governor Tennyson wrote in a dispatch to the Secretary of State for the Colonies: 'I herewith forward a flag with the new device upon it: the South Australian Shrike in the rising sun of the Commonwealth.'

Brolga

Piping Shrike or White-headed Magpie

Yellow Wattlebird

Today, the Piping Shrike's image is used by many government agencies as a corporate logo. It is a symbol of South Australia's resourcefulness and bravery.

TASMANIA

Yellow Wattlebird *Anthochaera paradoxa* Australia's biggest honeyeater. Only bird species endemic to Tasmania. Although Tasmania does not have any official fauna emblems, the Yellow Wattlebird, according to the Tasmanian Parliamentary Library, 'is generally acknowledged to be our most identifiable bird'.

VICTORIA

Helmeted Honeyeater *Lichenostomus melanops cassidix* One of four subspecies of the Yellow-tufted Honeyeater and highly endangered. Black, gold and olive, the Helmeted Honeyeater used to be found in the tributaries of the upper Yarra River and Western Port Bay drainages. Pure wild populations are now confined to just a small part of the Yarra River. Yellingboro State Wildlife Reserve was created to protect the few remaining colonies of 100 to 150 birds.

The Helmeted Honeyeater became Victoria's avian emblem in 1971.

Helmeted Honeyeater

Black Swan

WESTERN AUSTRALIA

Black Swan *Cygnus atratus* The Black Swan is firmly embedded in Western Australian culture, for the ancestors of the indigenous Nyungar people living in the southern and western parts of Western Australia believed they were Black Swans who became men.

Until Dutch explorers reached the 'Great South Land' in the 17th century, Europeans believed all swans were white.

Explorer and navigator Captain Willem de Vlamingh named the Swan River, where Perth now stands, in 1697 after seeing flocks of Black Swans in the estuary.

Around 130 years later, Captain James Stirling reported a flock of more than 500 Black Swans flying over the river.

The original European settlement in Western Australia was thus named the Swan River Settlement (or Colony), and the Black Swan became a symbol of the new community, appearing on government papers, banknotes, stamps and such like.

The Black Swan, which also appeared on the original state crest, became Western Australia's official bird emblem on 25 July 1973.

Today, it appears on the WA state flag and coat of arms.

AUSTRIA

Barn Swallow

Hirundo rustica

Austria 'shares' the Barn Swallow with Estonia, both countries having adopted this species as their national bird. However, the Barn Swallow does not appear to have a particularly high profile in Austria.

WHERE TO SEE The Barn Swallow is a common and widely distributed bird that is just as likely to be seen around houses as farm buildings. One of the joys of spring and summer is watching Barn Swallows effortlessly skimming low over the ground, vegetation or water, catching insects in mid-flight. Anyone fortunate enough to live near a nest site will appreciate the aerial agility of these birds as they swoop over, under and around obstacles, including admiring birdwatchers, with ease as they bring food back to their fledglings.

LEAST CONCERN

IUCN: There are estimated to be 190 million mature individuals globally. The Barn Swallow is the most familiar and widespread of all swallow species.

Size: Length 18 cm (7 in).

Description: Males of the nominate subspecies have blue-black upperparts and breast band, a dark red forehead and throat, off-white underparts and a deeply forked tail with streamers.

Diet: Mainly insects, including flies, ants, aphids and parasitic wasps, which are usually caught in flight. Eighty insect families have been recorded in this species' diet, according to volume 9 of the *Handbook of the Birds of the World*.

Reproduction: Builds a cup-like nest of mud and plant fibres which it 'cements' to the interior or exterior of a man-made structure such as a house, barn or other building. Clutch size is usually four or five eggs.

Range: Huge – 51.7 million sq km (19.10 million sq miles). Found in many parts of the world. Mainly a long distance migrant – British birds, for example, winter in South Africa.

Habitat: Wide variety of habitats, from open countryside, pastures, farmland and riparian areas to villages, towns and even cities in some places.

Far left: *Barn Swallows construct their nests mainly on or in any suitable man-made structures, especially barns and other suitable farm buildings, but also bridges and culverts.*

Left: *Two to seven speckled eggs are laid, although the average is four or five.*

Dozens of swallows – adults and youngsters alike – chattering away on telegraph wires is a common sight in many areas in summer and early autumn. Such gatherings often take place ahead of the long flight back to Africa where swallows spend the winter.

Cultural presence

One might be forgiven for thinking that an eagle of some sort is Austria's national bird, for a stylised single-headed eagle dominates the country's coat of arms. The single-headed eagle replaced the double-headed eagle that was used for heraldic purposes in the days of the Austro-Hungarian Empire.

However, it is the Barn Swallow, not an eagle, that is Austria's national bird. The Federal Ministry of Agriculture, Forestry, Environment and Water Management points out that although the heraldic bird of Austria is the eagle, the real national bird is the Barn Swallow, the latter having been chosen in 1960 by the International Council for Bird Preservation (now BirdLife International). The logo of BirdLife International features a Barn Swallow in flight.

Despite its low profile, one of the most notable cultural links with the Barn Swallow is that one of the waltzes composed by Josef Strauss (1827–1870) – *Dorfschwalben aus Osterreich* or Village Swallows from Austria – is named after this species. Josef Strauss, who was said by his famous brother Johann to be the more gifted of the two men, composed this waltz in 1864. His musical legacy is more than 300 original dances and marches and 500 arrangements of other composers' music.

Three Barn Swallows were depicted on one of the four bird stamps issued in Austria in 1953 by Austrian Post.

The Austrian coat of arms features a single-headed eagle; however it is not the national bird.

BAHAMAS

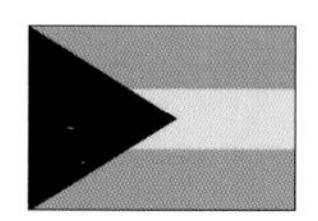

Caribbean Flamingo

WEST INDIAN FLAMINGO

Phoenicopterus ruber

The Caribbean Flamingo is the national bird of the Bahamas and its image is widely used. It forms an integral part of the country's coat of arms, with one of the flamingo's long legs supporting the central shield on which a radiant sun is depicted.

The Bahamas National Trust and the Trust's Discovery Club for children also use the flamingo on their logos.

The Caribbean Flamingo has appeared on many Bahamian stamps over the years, including a set of four issued in 2012 in association with the World Wide Fund for Nature (WWF) and a first-day cover issued in 1982 illustrating the life cycle of this species. It has also appeared on coins and banknotes.

WHERE TO SEE The world's largest breeding colony of Caribbean Flamingos is in the Bahamas National Trust-managed Inagua National Park on Great Inagua Island – a haven for birdwatchers.

'Marching' flamingos are a star attraction at Ardastra Gardens, Zoo and Conservation Centre in Nassau.

LEAST CONCERN

IUCN: There are between 260,000 and 330,000 individuals in the wild overall. The population trend is unknown.

Size: Largest of the western hemisphere flamingos, the biggest males standing nearly 145 cm (57 in) tall and weighing up to 4 kg (8.8 lb). Females are typically around 20 per cent smaller.

Description: The world's most colourful flamingo, with adults sporting bright orange-pink plumage and a distinctively shaped, pink, black-tipped bill. The Caribbean Flamingo used to be classified as a subspecies of the Greater Flamingo. Now both are considered by taxonomists to be full species.

Diet: Various organisms, including brine shrimps and other crustaceans, molluscs, larval aquatic insects, adult terrestrial insects, annelid worms, algae and diatoms. All flamingos are pink to a greater or lesser extent because of the carotenoid pigments found in their food.

Reproduction: A single egg is laid in a nest of hardened mud resembling an upturned cone.

Range: Found in the Caribbean, Mexico, South America and Galapagos Islands. This is the only flamingo species that does not share its range with other flamingos.

Habitat: Shallow, nutrient-rich bodies of water such as saltpans, saline lagoons and lakes, and alkaline lakes. Also frequents sewage-treatment facilities, dams, mudflats, estuaries and coastal waters.

Flamingos feed mainly on algae and tiny aquatic invertebrates which they filter from water using their highly specialised and oddly shaped bill.

Flamingos are very gregarious, often gathering in huge flocks numbering thousands or tens of thousands of birds.

Cultural presence

The Caribbean Flamingo's prospects looked decidedly poor in the late 19th and early 20th centuries. Large numbers of young birds were killed before they could fly, while many others were caught alive and sold to passing ships on which they died from lack of care.

At the inaugural AGM of the National Audubon Society in the USA in 1905, a plea was made to the government of the Bahamas for the Caribbean Flamingo to be legally protected. Later that year the Wildbirds (Protection) Act was passed. It was the first time in history that special protection had been proposed for flamingos – and then enshrined in law.

Despite this, flamingo numbers continued to fall, and by the 1950s it seemed as though the Caribbean Flamingo was heading for extinction, the number of wild birds having dwindled to only about 5,000 individuals. However, following the formation of the Society for the Protection of the Flamingo in 1951, and the Bahamas National Trust in 1959, the Caribbean Flamingo has made a comeback. Due to the work of the Bahamas National Trust, which is responsible for a Bahamas-wide network of 27 national parks covering an area of more than 404,700 hectares (1 million acres), Caribbean Flamingo numbers now exceed 50,000.

The flamingo depicted on the cartoon-like logo of the Bahamas National Trust's Discovery Club wears a green Trust hat 'to show pride' in what has been achieved down the decades in bringing the Caribbean Flamingo back from the brink of extinction.

Trained captive Caribbean Flamingos, which strut their stuff to the commands of a 'drill sergeant', are a popular tourist attraction at Ardastra Gardens, Zoo and Conservation Centre. Flamingos have been performing at Ardastra ever since the centre's founder discovered in the 1950s that his birds eventually responded to voice commands. Children especially enjoy the antics of the flamingos and often try to mimic them.

'Flamingo' is a commonly used place and name word in the Bahamas, examples being Flamingo Hills Resort and Marina on Cat Island, Flamingo Cay private cove on Andros Island and Flamingo Consultants.

Finally, at least one Bahamian beauty-show contestant has donned a flamingo costume to impress judges.

Right: *Flamingos are depicted on this 1971 Bahamian stamp.*

Far right: *The Caribbean Flamingo features on the logo of the Bahamas National Trust.*

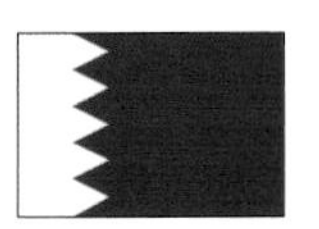

White-eared Bulbul

WHITE-CHEEKED BULBUL

Pycnonotus leucotis

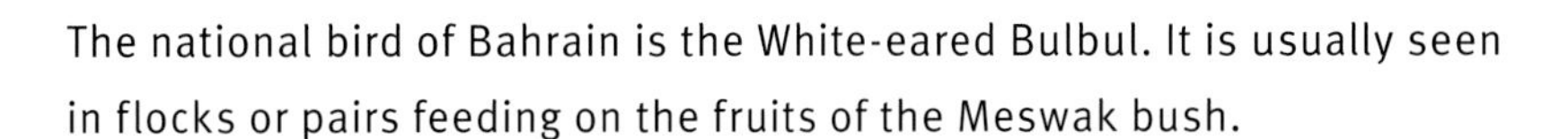

The national bird of Bahrain is the White-eared Bulbul. It is usually seen in flocks or pairs feeding on the fruits of the Meswak bush.

WHERE TO SEE There are many places to see the White-eared Bulbul, for this species can turn up in a range of contrasting habitats – anything from dry woodland, scrub, palm groves, orchards and gardens to mangroves, reedbeds and riverine shrubs.

LEAST CONCERN

IUCN: Unknown number of mature individuals, but said to be locally common. The population is thought to be declining.

Size: Length 17.5–19 cm (6.9–7.5 in).

Description: Nominate race has a black head and neck with a conspicuous white face patch, a blackish bill, light grey upperparts and even paler underparts, and an orange-yellow vent area. Sexes are similar. There are two subspecies of White-eared Bulbul – *Pycnonotus leucotis* and *P. l. mesopotamia*. The latter occurs in Bahrain.

Diet: Fruits, buds, nectar and invertebrates.

Reproduction: Usually three eggs are laid in a cup-shaped nest in a bush, some other shrub or small tree.

Range: Very large. Found in a number of countries in the Middle East and surrounding region. An introduced bird in Bahrain.

Habitat: Wide ranging. Dry savannahs and shrublands are major habitats, but this species is also found on arable land and in plantations, rural gardens, wetlands and deserts.

Cultural presence

Despite being Bahrain's national bird and a familiar species on the island of Bahrain, the White-eared Bulbul does not appear to have an especially high profile within the country. However, it has featured on a stamp in Bahrain.

This detailed illustration of a White-eared Bulbul appeared on a 1991 stamp in Bahrain.

BANGLADESH

Oriental Magpie-robin

Copsychus saularis

The Oriental Magpie-robin is the national bird of Bangladesh. Its image has been widely used in the country, and it has appeared on banknotes and stamps.

WHERE TO SEE This is a common and widely distributed bird in Bangladesh, so is likely to be seen almost anywhere.

LEAST CONCERN

IUCN: Although the global population of this species has yet to be determined, it has been described as a common to abundant bird.

Size: Length 19–21 cm (7.5–8.2 in).

Description: Pied bird, males having glossy blue-black upperparts and breast, a white belly, white wing bands, a long tail, and a black bill and legs. There are eight subspecies of Oriental Magpie-robin. It's the nominate race *C. s. saularis*, which occurs in Bangladesh.

Diet: Largely insects (from crickets and caterpillars to firebugs and flies), as well as assorted other invertebrates.

Reproduction: Two to five eggs are laid in an untidy nest often in a hole or other natural or man-made cavity.

Range: Very large – found in 17 Asian countries.

Habitat: Forests are especially important for this species. It is also found near rivers and streams, on arable land and in plantations, gardens and urban environments.

Cultural presence

Known locally as the Doyel or Doel, the Oriental Magpie-robin is the national bird of Bangladesh and is depicted in a landmark feature called Doyel Chatwar, which means Doyel Square, in Dhaka.

Left: *The Oriental Mapgie-robin appeared on this 1983 stamp – part of a Birds of Bangladesh series.*

Above: *This species was the official mascot for the 11th South Asian Games, held in Dhaka in 2010.*

Above: *The Doyel Chatwar statue is a tribute to the national bird of Bangladesh.*

BELARUS

White Stork

Ciconia ciconia

The White Stork is the national bird of Belarus. It was depicted on bird stamps in Belarus in 1995, 1996, 1998 and 2002.

White Stork coins of several dominations were minted in Belarus in 2009, the obverse side showing a stork's footprint on a grass background and the reverse showing a stork's head and a pair of nesting birds.

WHERE TO SEE Belarus is one of well over 100 countries in which the White Stork is a native bird. It breeds in Belarus and many other parts of Europe.

Areas in Belarus where the White Stork can be seen in spring and summer include the heavily forested Belovezhskaia Pushcha and the River Pripyat floodplain. A major complex of primeval forests, the Belovezhskaia Pushcha contains many trees that are 200 to 250 years old, while the Pripyat is internationally noted for its waterfowl during the spring migration period. BirdLife International reports that there were an estimated 220 to 250 breeding pairs of White Stork in Belovezhskaia Pushcha in 2001, and an estimated 300 to 500 breeding pairs on the Pripyat floodplain.

LEAST CONCERN

IUCN: There are 500,000 to 520,000 mature individuals globally. Numbers are increasing overall, although some populations are decreasing or stable.

Size: Length 110–115 cm (43.3–45.3 in). Very large bird.

Description: Long neck, bill and legs. Adults have a black and white body, and a bright red bill and legs. Unmistakeable, especially in flight.

Diet: Opportunistic. Takes a wide range of prey, including large insects, earthworms, small mammals, amphibians, reptiles and fish.

Reproduction: An average of four eggs are laid in a huge stick nest built either in a tree or on a man-made structure such as a roof, chimney, pylon, telegraph pole or electricity-supply pole (especially the latter). Following a decline in White Stork populations in some parts of western and southern Europe in recent decades, conservation efforts have included the building of man-made artificial nesting platforms in some countries.

Range: Extensive. Breeds mainly in Europe and winters mainly in Africa. European storks migrate south in late August in vast flocks, taking advantage of thermal updraughts, usually arriving in Africa by early October.

Habitat: Open lowland. Tends to prefer wetter areas in its breeding territory and drier terrain in its wintering quarters. Occurs in everything from damp or wet pastures, pools, marshy areas and arable fields to grasslands and savannah plains. Sometimes nests in towns and cities.

Many birds' nests are hidden from view among vegetation and difficult to spot unless one knows precisely where to look. But that is certainly not the case with White Storks. These large, charismatic birds boldly build very large nests on rooftops, chimneys, pylons, telegraph poles and other man-made structures, including specially erected nesting platforms, as well as more conventionally in trees and on cliffs. Made of large sticks, stork nests can be 2.5 metres or more deep.

Cultural presence

In a scientific research report entitled 'Preservation of White Stork (*Ciconia ciconia* L.) population in Belarusian Polessia', published in Minsk in 2000, resesarcher Irina Samusenko stated that the White Stork was the favourite bird of people in Belarus and that it played a major role in the country's cultural history.

She also added that the positive attitude of people living in rural areas had long guaranteed the conservation of this nationally revered species.

The White Stork was one of several birds illustrated in a BirdLife International series of stamps in Belarus in 2002.

Gallic Rooster

Gallus gallus

IUCN: Not evaluated

Size: Length 65–75 cm (25.6–29.5 in).

Description: A rooster is another name for a male chicken. It is a term more widely used in the USA than in the UK, where such a bird is commonly referred to as a cockerel or cock. Roosters are unmistakeable, sporting pillar-box red wattles and comb.

Diet: Given the chance, chickens forage for and eat a wide range of food, including seeds, grasses and insects. Many poultry keepers buy specially formulated balanced diets, the precise type used varying according to the ages of their birds. Chickens kept in back gardens and on smallholdings are also often given fresh fruits and vegetable scraps, as well as treats like mealworms.

Reproduction: Hens lay a variable number of eggs, according to their breed. Some lay a few dozen eggs annually, others more than 200.

Range: Does not occur naturally in the wild.

Habitat: Being a domesticated bird, the Gallic Rooster is common and widespread in a variety of situations – from gardens and smallholdings to mixed farms and specialist poultry farms. All domesticated chickens are believed to be descended from the tropical Red Junglefowl.

The so-called 'bold Rooster', or 'Walloon Rooster', was chosen as an emblem for the Walloon or Wallonia Region of Belgium (the southern, predominantly French-speaking part of the country) by the Walloon Assembly on 20 April 1913. Designed by Pierre Paulus, this male chicken or cockerel appears on the region's coat of arms and flag.

Some websites claim the Common Kestrel is the national bird of Belgium.

Cultural presence

Although various websites claim that the Common Kestrel *Falco tinnunculus* is Belgium's national bird, in fact this European nation does not have a generally recognised avian icon. However, the Walloon Region of Belgium has adopted the rooster as its emblem. It appears to be the same as the Gallic Rooster *Gallus gallus*. The Gallic Rooster – a domesticated rather than a wild bird – is the national bird of nearby France.

Walloon or Wallonia Region flags bearing the red image of a Gallic Rooster.

Keel-billed Toucan

Ramphastos sulfuratus

Known locally, and aptly, as the 'Bill Bird', the Keel-billed Toucan was adopted as Belize's national bird in 1981, when the former British Honduras became an independent country.

WHERE TO SEE This species can be found in many rainforest areas of Belize, including Chaa Creek Nature Reserve, where more than 300 bird species have been recorded.

LEAST CONCERN

IUCN: Global population estimated in 2008 to be anywhere between 50,000 and half-a-million individuals.

Size: Length 46–51 cm (19–20 in).

Description: An extremely colourful and easily recognisable bird, noted for its seemingly preposterously large, orange, blue and green, red-tipped bill. Maroon-glossed black head and back and yellow throat and breast, the lower part of which is edged with red. Females are smaller than males and have a shorter bill. There are two subspecies, *R. s. brevicarinatus* being the one found in Belize. It is usually smaller than the nominate race, with a slightly broader red breast band.

Diet: A variety of fruits, as well as invertebrates (for example beetles, spiders and ants), snakes, lizards and probably birds' eggs.

Reproduction: One to four eggs are laid in a natural tree cavity roughly 45 cm (17.7 in) to 2 m (6.6 ft) deep. If breeding is successful, the nest is used again the following year.

Range: Very large. Native to nine Central and South American countries: Belize, Colombia, Costa Rica, Guatemala, Honduras, Mexico, Nicaragua, Panama and Venezuela.

Habitat: Mainly damp lowland forests, but also overgrown plantations and open pastures.

Cultural presence

Native tribes and early European colonists were not only aware of toucans of various kinds but also caught them for food and to keep as pets. The colourful feathers of these birds were highly prized by Amerindians, and their body parts, particularly their bills, were used in traditional medicine.

In recent times, the Keel-billed Toucan has often been depicted on stamps in Belize. In 1981, for example, it was featured on an independence stamp, in 1986 on one of four different toucan stamps and in 2006 on a 25th anniversary of independence stamp.

The Keel-billed Toucan was featured on a 10c stamp in 1986.

Fruits and berries of various kinds, including bananas, are the staple diet of the Keel-billed Toucan, which is mainly a canopy-feeding bird in Belize.

BERMUDA

Bermuda Petrel

CAHOW

Pterodroma cahow

● **ENDANGERED**

IUCN: Extremely rare, long-lived seabird related to albatrosses and shearwaters.

Size: Length 38 cm (15 in); wingspan 1 m (3.3 ft).

Description: Long-winged, medium-sized, fast-flying petrel. Mainly white underparts contrast sharply with brownish-grey upperparts. Brownish-black cowl, black bill, and pink legs and feet.

Diet: Squid, krill and anchovies. Adults fly thousands of kilometres across the ocean during the breeding season to catch food for their offspring.

Reproduction: A single egg is laid in January in a natural or artificial burrow.

Range: An oceanic wanderer. Spends most of its life well away from land in the North Atlantic. Nests only in Bermuda, to which it returns in November for courtship, nest-building and breeding.

Habitat: Pelagic outside the breeding season. Nonsuch Island and several rocky islets in Bermuda during the breeding season.

The Cahow, as this bird is known in Bermuda, has been featured on stamps, coins and banknotes, and is the logo of the Bermuda Government's Department of Parks. It was officially declared Bermuda's national bird in 2003.

WHERE TO SEE Cahows are nocturnal over the breeding islands, only visiting their nest burrows at night. It is possible to see them in Bermuda in late afternoon in November, and between January and April as they fly offshore. The best location is Cooper's Point, on Cooper's Island Nature Reserve, where the birds can be viewed outside the reef line using a telescope. However, to get close-up views of Cahows or to photograph them, it is necessary to go on one of the Bermuda Audubon Society's pelagic birdwatching trips, mainly in November. Such trips are weather dependent.

A Bermuda Petrel over its breeding island. Strictly nocturnal on land, this species locates its burrow in darkness by calling to its mate and landing nearby once a response has been received.

Cultural presence

A fully-fledged Bermuda Petrel chick exercising its wings.

As a Spanish ship sheltered from a storm in a cove on uninhabited Bermuda in 1603, hundreds of eerily calling petrels suddenly surrounded the vessel. Unfamiliar with these birds and already frightened by the tempest, the superstitious sailors believed them to be devils. Unfortunately for the petrels, it was soon discovered that they were easy to catch and rather good to eat. As a consequence, by the 1620s the Cahow was believed to be extinct as a result of hunting by people and predation by introduced pigs, rats, cats and dogs. A bird that nested only on the islands of Bermuda and thought to have originally numbered more than half a million individuals had gone the way of the Dodo – or so it seemed.

Centuries later, in 1906, a small, unidentified petrel found on Bermuda's Gurnet Rock by L. L. Mowbray proved to be a Cahow. When the bird was compared with the bones and subfossil remains of the original Cahow found in caves and crevices, 'it was proved beyond doubt', wrote world-renowned conservationist Dr David Wingate, that the 'new' bird was not in fact new at all, but the species long thought to have been extinct. In a paper entitled 'The Fabled Cahow', Dr Wingate said that 'A handful of survivors had evidently been coming and going silently in the night, incubating their single eggs and raising their young unseen in the remote rock burrows of the outer islands and keeping the species alive, unknown to man.'

Other Cahows turned up in 1935 and 1941, one having flown into a lighthouse and the other into a telephone wire. The big breakthrough came in 1951, when a small breeding population of the petrels was discovered on four offshore islets.

As a result of Bermuda's long-running Cahow Recovery Programme and the sterling work of now-retired Dr Wingate, the number of Cahow pairs has grown from a paltry and precarious 17 in 1962 to just over 100 today.

Nearly 75 per cent of all Cahows now nest in 'des-res' concrete burrows created for them by the Government of Bermuda to overcome the problem of insufficient soil on most of the nesting islands. The artificial burrows have been designed in such a way as to let Cahows in, but keep out White-tailed Tropicbirds (Longtails) *Phaethon lepturus*, which compete for nesting space.

Nonsuch Island has been restored as a 'living museum' and is potentially capable of supporting thousands of pairs of Cahows. Predators have been removed and there is enough soil on the island into which Cahows can burrow naturally. The habitat now closely resembles what early settlers described Cahows as having nested in.

Between 2004 and 2008, a total of 105 near-fledged Cahow chicks were translocated from their nesting islets to Nonsuch Island, where they were fed squid and fish, fitted with tags and monitored until they flew out to sea.

No fewer than 102 of these chicks successfully fledged. In 2008 the first of these now fully grown Cahows returned to Nonsuch Island prospecting for nest sites. Tags identified four of these birds as having left Nonsuch in 2005. A significant milestone was achieved when a pair of birds produced its first chick in 2009 – the first Cahow to hatch on Nonsuch Island in nearly 400 years.

In 2012 the number of translocated Cahows that had returned to Nonsuch Island had risen to 26, with 10 breeding pairs producing a total of seven fledged chicks. By the 2013 breeding season the number of returned Cahows had risen to 45, with a total of 12 breeding pairs at the new Nonsuch Colony, all of which were incubating eggs.

Cahow Recovery Programme manager Jeremy Madeiros says that the long-term aim is to increase the nesting population to a minimum of 1,000 breeding pairs so that the conservation status can be improved from Critically Endangered to Threatened. Specific conservation objectives include preventing nest-site competition with White-tailed Tropicbirds by installing wooden 'baffler plates' at the entrances to all Cahow burrows to prevent tropicbirds from entering them, and regularly monitoring all nesting islands for the presence of rats. Where rats are detected or suspected to be present, they will be eradicated using anti-coagulant rodenticides. Madeiros says that the government will continue to provide additional nest burrows on all appropriate nesting islands and wherever new colonies are being established.

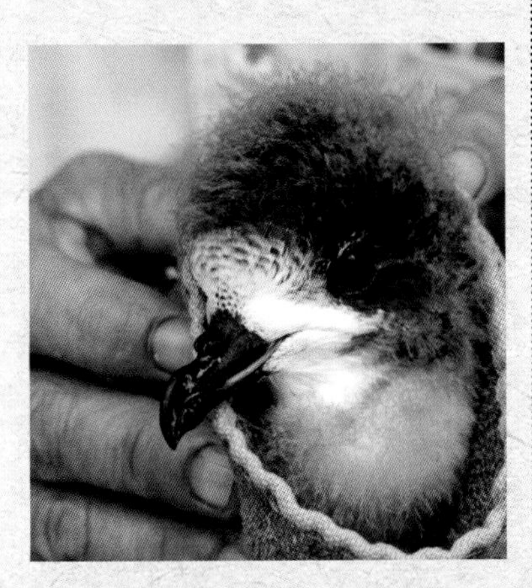

Bermuda Petrels are long-lived birds and may reach the ripe old avian age of more than 40 years. Pairs stay together for many years – possibly for the whole of their breeding lives.

Another objective is to 'establish new nesting colonies of Cahows on larger, more elevated islands free of mammal predators and safer from hurricane erosion and which have the potential of supporting larger populations of the birds. This is already being achieved on Nonsuch Island and is planned for the future on Southampton Island.' Yet more, it is hoped, will be learned about the biology of the Cahow, 'through an on-going banding programme initiated in 2002, and through developmental studies of chicks and adults'.

Studies of the Cahow's oceanic range using geolocators attached to individual birds 'have already been successful in recording foraging areas and migration routes used by Cahows – when they are foraging for their chicks during the breeding season and when they are travelling during the summer, non-breeding season'.

In recent years the Cahow, Nonsuch Island and the translocation project in particular have been the subject of three documentaries: Bermuda's *Treasure Island* by Deirdre Brennan, *Rare Bird* by Lucinda Spurling and *Higher Ground* by LookTV.

Andrew Dobson, President of the Bermuda Audubon Society, says that the Cahow 'has never received the attention it deserves. It was only a few years ago that it was declared the national bird. Prior to that, everyone assumed it was the flashy Longtail which is featured on just about everything!' Although the White-tailed Tropicbird is the Audubon Society's logo, Andrew says, 'we highlight the Cahow at every opportunity.'

The Bermuda Petrel was depicted on this $5 stamp in 1978 – one of five in a wildlife series.

***Source:** reproduced by permission of the Bermuda Philatelic Bureau*

Common Raven

Corvus corax

The Common Raven is highly revered in Bhutan, not least because it represents Jarog Dongchen, a powerful guardian deity. Indeed, the Common Raven, which is Bhutan's official national bird, is known in the local language as Jarog. A key symbol of royal authority in Bhutan is the Raven Crown, or Uzha Jarog Dongchen – not just a raven-topped hat but, more importantly, the personification of the raven-faced Jarog Dongchen.

WHERE TO SEE Masters of the wind, Common Ravens are most likely to be seen first on the wing – twisting and turning, ducking and diving, and performing other aerial manoeuvres. In Bhutan, these birds often nest in monasteries and dzongs (fortresses). According to one website, ravens can be seen in western Bhutan at Cherithang, Lingshi, Damthang and Chebesa, at Pegula, Dur and Bumthang in central parts of the country and in the Singye Dzong area of eastern Bhutan. These large black crows move to lower altitudes in winter.

LEAST CONCERN

IUCN: The global population was estimated in 2004 to be around 16 million individuals.

Size: Length 58–69 cm (22.8–27.1 in); weight 585–2,000 g (20.6–70.5 oz). The world's biggest crow.

Description: The quintessential big black bird. The nominate race is blue-purple or green-glossed black all over. Even the legs and bill of this species are black. Shaggy throat and diamond-shaped tail. Sexes are similar, although males are much bigger than females. There are eleven subspecies, the one found in Bhutan being *C. c. tibetanus*, which is very big and extremely glossy.

Diet: The Common Raven is a classic example of an opportunistic scavenger. It will eat virtually any carrion it comes across, including road kills, slaughterhouse offal, placenta and entrails, as well as eggs, young birds, mammals, amphibians, snakes and fish, and a range of invertebrates such as slugs, beetles and worms. This species is also an omnivore, eating a variety of plant material like fruit and grain.

Reproduction: Usually four to six eggs are laid in a large, platform-like stick nest, which can be in a tree, on a cliff ledge or on a man-made structure of some sort, such as a tower, derrick or bridge.

Range: Huge. Native to some 80 countries, including Bhutan.

Habitat: Common Ravens are generalists in that they occur in a wide range of contrasting habitats such as forests, cliffs and other coastal areas, mountains, and desert, steppe and tundra regions.

Above: *Common Ravens often nest in monasteries and dzongs in Bhutan – not unlike the craggy cliff faces they are more commonly associated with in the wild.*

Right: *Although ravens are actually crows, they are so big they are often mistaken in flight for birds of prey.*

Cultural presence

The image of a raven has appeared in one form or another on stamps, coins and banknotes in Bhutan, and formed part of the logo for the country's coronation and centenary celebrations in 2008.

The Raven Crown was also used in 2008 during the coronation of Oxford-educated Jigme Khesar Namgyel Wangchuck as the fifth Dragon King (Druk Gyalpo) of the Kingdom of Bhutan. It was adopted as a symbol of royal authority and power by the first Wangchuck Dynasty monarch, Ugyen Wangchuck (1862–1926). The original version of the crown was designed as a battle helmet and worn by Ugyen's father, the Black Regent.

An angel carrying a Raven Crown was depicted on a 100-ngultrum note issued by the Royal Monetary Authority of Bhutan on 13 October 2011 to mark the marriage of King Jigme Khesar Namgyel Wangchuck and Queen Jetsun Pema.

Killing a raven was once a capital offence in Bhutan.

Left: *The Common Raven was one of four national symbols illustrated on stamps in Bhutan in 2005.*

Right: *The Raven Crown of the Dragon Kings of Bhutan – a soft hat, the top of which has been designed to resemble a raven's head.*

BOLIVIA

Andean Condor

Vultur gryphus

The Andean Condor is featured prominently on Bolivia's coat of arms, where it represents liberty. The coat of arms, in turn, forms part of the country's national flag.

The Andean Condor has appeared on various Bolivian stamps (including the 1925 independence centenary issue and 2007 Birds of La Paz set), is the subject of vibrant, highly stylised paintings by Bolivian artist Roberto Mamani Mamani, and is the logo of regional airline Aerolineas Argentinas, which flies to various South American countries, including Bolivia.

WHERE TO SEE The Apolobamba Mountains of the Greater Madidi-Tambopata Landscape in north-eastern Bolivia are a regional stronghold for the Andean Condor, with 80 to 150 individuals present in this area, according to the Wildlife Conservation Society in Bolivia (WCS Bolivia).

NEAR THREATENED

IUCN: The overall population is estimated to be 10,000 mature individuals and thought to be declining.

Size: Length 100–130 cm (39.4–51.2 in). Wingspan up to 3.2 m (10.5 ft). The world's biggest raptor and South America's largest flying bird.

Description: Adults of both sexes have a bare head, predominantly black plumage and a conspicuous white ruff. Males, which have yellow eyes, are also endowed with a large comb and neck wattle. Females lack the comb and wattle and have red eyes.

Diet: Mainly carrion. Scavenges on the carcasses of mammals, including tapir, deer, rodents and domestic livestock.

Reproduction: One egg is laid on a cliff ledge or in a small cave. The chick is nurtured by its parents for up to two years and takes five to eight years to reach maturity.

Range: Native to eight South American countries (Argentina, Bolivia, Chile, Colombia, Ecuador, Paraguay, Peru and Venezuela), and breeds in almost all of them including Colombia.

Habitat: Usually found well away from people among the peaks and valleys of the High Andes.

Far left: *Bulky, heavy and with a huge wingspan, the Andean Condor is a seriously impressive bird.*

Left: *The Andean Condor's featherless head is superbly adapted for this species' scavenging way of life, for this bird often inserts its naked head inside the carcasses of animals to tear off chunks of rotting flesh. Feathers would quickly become matted with blood and gore.*

Cultural presence

The Andean Condor is deeply embedded in Bolivia's culture. Carved stone figures with condor heads have been found at the Sun Gate of Tiwanaku or Tihuanaco – the megalithic ruins of an ancient civilisation that once flourished on the shores of Lake Titicaca. The precise origins, nature and age of this pre-Hispanic culture have long been the subject of controversy. The Seventh Summit of the Eagle and the Condor – a gathering of Native American spiritual guides – was held at Tiwanaku in 2009.

Condortrekkers is an organization based in Sucre, Bolivia, specialising in trekking and city tours, while *The Blood of the Condor* is the name of a 1969 Bolivian film, directed by Jorge Sanjine,s about a poor community that rises up against exploitation.

Created in 1921, the National Order of the Condor of the Andes is awarded in recognition of exceptional civil or military achievements by Bolivians and other nationals. The town of Laja is decorated with miniature condor heads, fruit and small fish made from bread each February to celebrate incoming mayors, commissioners and the like.

'The Andean Condor is considered sacred by Andean people, but at the same time is also perceived as the cause of livestock loss by many Andean communities,' said Rob Wallace, Director of the Greater-Tambopata Landscape Conservation Programme of the Wildlife Conservation Society in La Paz, Bolivia. 'There is some anecdotal evidence to support this, although the perception is probably exaggerated.

Above: *The Andean Condor appeared on this Bolivian bird stamp in 2007.*

Above: *A protester in La Paz, Bolivia, 2011, dressed as an Andean Condor for the Day of the Right to Regain Access to the Sea – an ongoing dispute with Chile over access to a Pacific-coastal stretch Bolivia lost in war many years ago.*

Left: *The Andean Condor forms a central part of Bolivia's coat of arms which is used, as shown here, on the country's national flag.*

Reverence is the best thing the Andean Condor has going for it.' He added that the Condor's image is 'very frequent' in Bolivia, 'and the fact that political leaders are called *mallku*'s, meaning High Andean Spirit or Condor, is something that must form part of an outreach and communication campaign for the conservation of the Condor'.

BOTSWANA

Lilac-breasted Roller

Coracias caudatus

Kori Bustard

Ardeotis kori

Botswana does not have an official national bird, according to both the Department of Environmental Affairs in Gaborone and BirdLife Botswana, yet many people believe it does. The species most commonly associated with Botswana is the beautiful Lilac-breasted Roller.

A number of websites erroneously state that the roller is indeed Botswana's national bird, but a quite different species, the Kori Bustard, is also cited by others. Both species are profiled here.

Dr Kabelo Senyatso, Director of BirdLife Botswana, says there are several unofficial contenders for national bird, notably Kori Bustard and Lilac-breasted Roller. The Cattle Egret (*Bubulcus ibis*) is another favoured speces. He added that BirdLife Botswana tried to persuade the Office of the President and Ministry of Environment, Wildlife and Tourism (MEWT) to become 'interested in a process to get Botswana to nominate a national bird. At the time, we couldn't proceed as MEWT cited budget constraints in organising such a consultative process, but I'm optimistic this is something they'll reconsider in the near future.'

Even though the Lilac-breasted Roller has no official status in Botswana, it is one of the species that most birdwatchers visiting the country for the first time want to see. In that respect, at least, it is a great ambassador for Botswana.

WHERE TO SEE It is widely distributed and prefers open woodland and savanna (as before).

LILAC-BREASTED ROLLER

LEAST CONCERN

IUCN: The size of the overall population is unknown, but it appears to be stable.

Size: Length 28–30 cm (11–11.8 in) plus long tail streamers.

Description: A veritable rainbow of a bird that, like Joseph, wears a 'coat' of many colours – white, green, brown, azure blue, dark blue, lilac and orange-pink. Sexes are alike. There are two subspecies of Lilac-breasted Roller – *Coracias caudatus lorti* and *C.c.caudatus*. It is the latter (the nominate race) which occurs in Botswana.

Diet: A variety of invertebrates, including locusts, crickets, grasshoppers, beetles, butterflies, caterpillars, scorpions and spiders, as well as frogs, lizards and birds.

Reproduction: Two to four eggs are laid on a grass pad in a tree cavity.

Range: Very large. Native to 19 African countries, including Botswana.

Habitat: Open Acacia woodland, grassland, farmland, wildlife and game areas, and such like. Often seen perching on fenceposts and telegraph wires.

The multicoloured Lilac-breasted Roller is one of Botswana's most beautiful and familiar birds.

KORI BUSTARD

NEAR THREATENED

IUCN: Unknown number of mature individuals, but said to still be common where undisturbed. Unfortunately, the Kori Bustard is not only illegally hunted for food but also trapped for the live-bird trade.

Size: The world's heaviest flying bird. Males are around 120 cm (47.2 in) in length and weigh 10.9–19 kg (24–41.8 lb), while the considerably smaller females are about 90 cm (35.4 cm) in length and tip the scales at a more modest but still large 5.9 kg (13 lb).

Description: Largely a brown, black, buff and white bird with a black and grey crown and long, pale-coloured legs. Despite its size, the Kori Bustard blends remarkably well with its surroundings. Although the sexes differ greatly in size, their plumage is largely the same. There are two subspecies of Kori Bustard – *Ardeotis kori struthiunculus* and *A . k. kori*. It is the latter (the nominate race) which occurs in Botswana.

Food: Omnivorous. Eats a wide range of food, from berries, bulbs and seeds to snails, insects, birds' eggs and nestlings, rodents, lizards and snakes.

Reproduction: Usually two eggs are laid in a shallow scrape in the ground.

Range: Very large. Native to 15 African countries, including Botswana.

Habitat: Mainly grass or scrub-covered open country.

Below: *Although Kori Bustards are much more likely to be seen strutting their stuff slowly and purposefully across grassland, they are also powerful fliers.*

The Kori Bustard is nowhere near as colourful as the Lilac-breasted Roller, but it is eye-catching because of its large size and stately appearance. A detailed study of the ecological needs and population dynamics of the bustard has been undertaken by Dr Senyatso of Birdlife Botswana to provide data that will hopefully lead to the improved management and long-term survival of this species.

Dr Senyatso says a public awareness and education campaign for the Kori Bustard would "go a long way" towards increasing appreciation for this species which is an "integral part of our culture".

If the Kori Bustard were to be become Botswana's national bird, such a designation would boost its status and ensure Botswana remained a stronghold for this bird.

WHERE TO SEE Although the Kori Bustard is a more elusive bird, it is not especially difficult to find. Populations of Kori Bustard still exist in Kgalagadi Transfrontier Park, Central Kalahari Game Reserve and Chobe National Park.

Cultural presence

Being a big bird has contributed to the Kori Bustard's downfall, for it is a highly prized target for illegal hunters. As BirdLife Botswana points out, just one bird can fill a larder for days. The Kori Bustard has long been regarded as a royal bird whose meat could be eaten only by tribal chiefs. Overhunting led to a significant drop in their population in Botswana, but conservation laws have helped the species to recover its numbers.

Another major threat to the long-term survival of this species is the live capture of individuals for illegal trading with neighbouring countries. BirdLife Botswana says that although the Kori Bustard is not globally threatened, 'its future in Botswana is cause for concern. It seems fated to an existence in protected areas alone.'

The relatively common Lilac-breasted Roller is often one of the first birds seen by tourists and birdwatchers on safari. It favours perching on posts, poles, dead branches and other vantage points, usually permitting good views. They are fiercely territorial and are found in abundant numbers in the Okavango region.

November to March is the best time for birdwatching in Botswana when the migrants are around.

Botswana has issued a wealth of highly collectable bird stamps over the years, but the Lilac-Breasted Roller, rather surprisingly, appears never to have been featured on any of them. The roller is an extremely photogenic bird and is thus often pictured in magazines and on websites.

Kori Bustards lay large eggs – usually two. They range in size from 81 to 86 mm by 58 to 61 mm and in weight from 121 to 178 g.

Typical Kori Bustard habitat.

Rufous-bellied Thrush

Turdus rufiventris

LEAST CONCERN

IUCN: Reported to be common, although the overall population is unknown.

Size: Length 23–25 cm (9–9.8 in).

Description: Brown head and upperparts, a streaked pale throat, a buff-brown breast and a rufous-orange belly. Sexes are similar.

Diet: Mainly fruits, especially berries, but also earthworms, spiders, insects and other invertebrates.

Reproduction: Usually three eggs are laid in a cup-shaped nest in a tree or shrub.

Range: Found in five South American countries – Argentina, Bolivia, Brazil, Paraguay and Uruguay.

Habitat: As varied as woods, thickets, forest edges, farmland and urban environments. Frequents gardens in Rio de Janeiro city.

Known locally as the *Sabia-laranjeira*, or simply *Sabia*. it took many years of lobbying before the Rufous-bellied Thrush was officially chosen as Brazil's national avian icon in 2002.

WHERE TO SEE An adaptable species, the Rufous-bellied Thrush is capable of turning up almost anywhere.

Cultural presence

The Rufous-bellied Thrush was one of a number of birds featured on Brazilian stamps in 1994.

***Source:** Brazilian Post and Telegraph Company*

It took many years for the Rufous-bellied Thrush, known locally as the *Sabia-laranjeire* or *Sabia*, to be officially chosen as the national bird of Brazil.

The Brazilian Wildlife Preservation Association proposed as long ago as 1968 that the Sabia should become Brazil's national bird. The first formal attempt to declare it as such was made in 1987. Despite ending in failure, the campaign attracted widespread public support. Much to the delight of this species' supporters, the second attempt, made in 2002, was successful. On 4 October of that year, by presidential decree, the Rufous-bellied Thrush at last became Brazil's official national bird. The Rufous-bellied Thrush was featured on the logo of the 2013 FIFA Confederations Cup hosted by Brazil. It is also the name of a Brazilian arts magazine, the symbol of which incorporates a small silhouette of the thrush.

A noted songster, the Rufous-bellied Thrush has been immortalised in various Brazilian poems and songs.

The Southern Lapwing *Vanellus chilensis* is the symbol of the Brazilian state of Rio Grande do Sul.

Zenaida Dove

Zenaida aurita

The Zenaida Dove has been national bird of the British Virgin Islands since 2012. Frequently, the Zenaida Dove is confused with similar species such as the Mourning Dove *Zenaida macroura* or an unspecified 'turtle-dove'.

WHERE TO SEE Native to 16 countries or territories, including the British Virgin Islands, the Zenaida Dove, which coos mournfully, is a common and widely distributed Caribbean bird.

LEAST CONCERN

IUCN: The overall population is very large and estimated to comprise anything from half a million to five million mature individuals.

Size: Length 28–30.5 cm (11–12 in).

Description: A small, attractively marked dove. Adult males are a smorgasbord of cinnamon, buff-brown, rufous-brown and similar hues. They also have an iridescent purple neck patch, blackish-brown primary feathers, a black bill and red legs and feet.

Diet: Fruits and seeds of wide-ranging plants.

Reproduction: Two eggs are usually laid. Nests mainly in trees and shrubs and on the ground on islands where there are few or no predators.

Range: Found throughout the West Indies and part of Mexico (north coast of the Yucatan Peninsula and Cozumel Island).

Habitat: Varied, including open woodland, second-growth forest, scrub and mangrove areas.

Cultural presence

Thought to have been sighted first by explorer Christopher Columbus in 1493, the British Virgin Islands are steeped in history and culture. To instil a sense of national pride and identity among young people, the government of the British Virgin Islands adopted a territorial song and dress in 2012, with the national bird being incorporated into the latter. Acting Director of Culture Mrs Brenda Lettsome-Tye remarked at the time:

'We have many aspects of our culture represented in the "Virgin Islands Print" that makes the fabric for the "Territorial Dress". It depicts the turtle dove, oleander, hibiscus, soursop, sugar apple and Virgin Islands sloop, along with the islands of the Virgin Islands. This print is a clear representation of those things that are synonymous with us as Virgin Islanders.'

The Zenaida Dove appeared on 5-cent stamps issued in 1973 and on 5-dollar stamps released in 1985 as part of a birds of the British Virgin Islands set. This species has also appeared on some coins.

The Zenaida Dove was depicted on this $5 stamp in 1985 as part of the Birds of the British Virgin Islands series.

Grey Peacock-pheasant

Polyplectron bicalcaratum

The Grey Peacock-pheasant, national bird of Burma, is known within the country as the *Chinquis*.

WHERE TO SEE This is a secretive, wary and therefore difficult-to-spot bird. Its presence is most likely to be betrayed by the male's shrill whistle and raucous call.

Cultural presence

The bird figureheads on the Karaweik Royal Barge, moored on the Kandawgy Lake in Yangon, resemble the Grey Peacock-pheasant.

The Karaweik Royal Barge, designed by U Ngwe Hlaing, is now a floating restaurant.

LEAST CONCERN

IUCN: The global population has yet to be ascertained. It is thought that this species is declining as a result of habitat loss and degradation, and in some areas overexploitation.

Size: Length 56–76 cm (22–30 in) males, 48–55 cm (18.9–21.6 in) females.

Description: Large, elegant, long-tailed, grey-brown pheasant. Upperparts adorned with eye-like 'spots', or ocelli, some of which are iridescent blue-green. Males have a prominent forwards-hanging crest. Females are darker, duller and smaller. There are five subspecies of Grey Peacock-pheasant, *Polyplectron bicalcaratum*. The nominate subspecies, *P. b. bicalcaratum*, is found in western and southern Myanmar (and in certain other countries). Volume 2 of the *Handbook of the Birds of the World* says *P. b. bakeri* may also occur in Myanmar.

Diet: Forages quietly and slowly for berries and other fruits, seeds, insects (especially termites), snails and other small creatures.

Reproduction: Usually two eggs are laid in a scrape or hollow within thick vegetation such as a bamboo clump or a tangle of bushes.

Range: Very large. Native to nine Asian countries/territories (Bangladesh, Bhutan, Cambodia, mainland China, India, Laos, Burma, Thailand and Vietnam). Common to rare depending on the area.

Habitat: Tropical evergreen and semi-evergreen forests with dense undergrowth.

CAMBODIA

Giant Ibis

Pseudibis gigantea

● **CRITICALLY ENDANGERED**

IUCN: There are believed to be only 200 mature individuals left in the wild as a result of hunting, drainage of wetlands for farming, deforestation and disturbance. The population appears to have declined rapidly during the past three generations and is continuing to do so.

BirdLife International reports that the Giant Ibis appears to be very sensitive to human disturbance, especially during the dry season when birds concentrate around available waterholes.

Size: Length 102–106 cm (40.2–41.7 in).

Description: A really big and bulky wading bird with a long, decurved bill. Easily the world's biggest ibis. Adults have a dark grey-brown plumage and red legs. The striking thing about the legs is that they are not especially long in relation to the body size.

Diet: Invertebrates, crustaceans, small amphibians and reptiles.

Reproduction: Nests in large trees, usually well away from human habitation. Two eggs are generally laid per clutch.

Range: Historically found in southern Vietnam and south-east and peninsular Thailand, the Giant Ibis is now restricted mainly to northern Cambodia where, reports BirdLife International, it is 'still fairly widespread but extremely rare'.

The Giant Ibis is native to Cambodia, Laos, Thailand and Vietnam.

Habitat: Seasonal pools, marshes and water meadows, as well as wide rivers, in dry lowland forests.

Known locally in Khmer as *Tror Yorng*, *Kangor Yak* and *Aov Loeuk*, the Giant Ibis is officially the national bird of the Kingdom of Cambodia. Hunting, which is one of the key threats facing this species, was outlawed by the Ministry of Agriculture, Forestry and Fisheries in 1994.

On 1 May 2013, Giant Ibis Transport signed an agreement with BirdLife International to provide US $51,000 to help conserve the Giant Ibis in Cambodia over a three-year period and, in so doing, became the first BirdLife Species Champion in that country. BirdLife Species Champions are a growing community of companies, institutions and individuals supporting conservation measures aimed at preventing bird extinctions.

WHERE TO SEE The Giant Ibis can be seen at Tmatboey village in Kulen Promtep Wildlife Sanctuary in the Northern Plains of Cambodia – the country's biggest protected area. Some areas of high density still exist, according to BirdLife International, such as in Preah Vihear Protected Forest and probably also in Siem Pang.

Cultural presence

Following the discovery in 2000 by the Wildlife Conservation Society (WCS) of the continued existence of the Giant Ibis in Cambodia's Northern Plains and the considerable interest shown by birdwatchers and other people, the WCS set up the Tmatboey Ibis Ecotourism Project in 2005. The aim is to use ibises – not just the Giant Ibises – as flagship birds for the protection of all globally threatened large waterbirds in Tmatboey through community-based tourism directly linking income to long-term conservation.

The agreement between the community, wildlife sanctuary and WCS stipulates that tourism payments to villagers for accommodation, meals, transport, guiding and the like are dependent upon them managing habitats, protecting key species and exercising a no-hunting policy. A community conservation contribution is also included in the price of all tourist visits to Tmatboey. The community management committee can choose how to spend this money. Typically, it is used to improve living conditions in the village, for example by building wells and schools, buying medical supplies and improving access roads.

As a result of the Tmatboey Ibis Ecotourism Project, wildlife populations are increasing and tourism enquiries are growing annually by about 30 per cent. Tmatboey was selected as the best community-based tourism site in Asia by the WildAsia Responsible Tourism Awards in 2007.

The subject of songs and traditional stories, the Giant Ibis is described by the Tmatboey Ibis Tourism Site as 'a near-mythical species for all birdwatchers, naturalists and conservationists'.

Cambodian farmers receive a premium price for their rice if they agree to abide by agreements conserving rare waterbirds and other wildlife in protected areas. Implemented by WCS Cambodia in partnership with the Ministry of Environment and Forestry Administration, the Wildlife-Friendly Ibis Rice Project bought more than 30 tons of Malis (fragrant) rice from farmers in participating villages between December 2008 and January 2009 at nearly double the price initially offered by local middlemen. After milling, 20 tons of rice was marketed in a range of outlets in Siem Reap and Phnom Penh, including upmarket hotels, restaurants and supermarkets. This scheme has continued to grow annually. In 2011, 147 tons of Malis rice were bought from 157 participating farmers in seven villages. A green silhouette of this bird forms part of the logo of Ibis Rice.

In 2003 the WCS initiated a Bird Nest Protection Programme under which local people living in the Northern Plains are offered a reward for reporting any nests they find of Giant Ibis and other globally threatened species, and are employed to monitor and protect birds until the chicks fledge. If chicks fledge successfully, the daily wage that is paid to protect the nest is doubled as a reward.

The Giant Ibis was depicted on a 2006 stamp bird stamp in Cambodia.

The Giant Ibis is mentioned in Khmer songs and traditional tales.

It is reported that the blood of this bird was once used to treat malaria in remote Cambodian villages.

The Giant Ibis and its morning calls are said to be widely liked, especially by farmers.

The Giant Ibis has also been depicted in public-awareness literature in Cambodia and Laos aimed at reducing the hunting of big waterbirds, as well as on at least one Cambodian stamp. It is also the logo for Giant Ibis Transport, a new upmarket bus service for local and international passengers.

Cayman Parrot

Amazona leucocephala hesterna/caymanensis

The Grand Cayman Parrot *A. l. caymanensis* and the Cayman Brac Parrot *A. l. hesterna* are the Cayman Islands' national birds, the two subspecies of the Near Threatened Cuban Amazon having been formally adopted as such by the government and people on Earth Day 1996.

WHERE TO SEE Queen Elizabeth II Botanic Park, Mastic Reserve and Trail and Central Mangrove Wetland in Grand Cayman are good viewing sites.

NEAR THREATENED

IUCN: The Grand Cayman and Cayman Brac are two subspecies of the Cuban Amazon. Figures from the Cayman Islands Department of Environment show population sizes are approximately 573 on Cayman Brac (2012 figure) and 4,308 on Grand Cayman (2011 figure). The Cayman Brac Parrot, which has a range of only 36.3 sq km (14 sq miles), is one of the world's rarest Amazon parrots.

Size: Length 32 cm (12.6 in). The Cayman Brac Parrot is smaller than the Grand Cayman Parrot.

Description: The Grand Cayman Parrot is mainly green with bright blue areas under its wings and tail, and a reddish face. Males of this subspecies are bigger than the females and more brightly coloured. The Cayman Brac Parrot has a maroon belly and more black on its feathers than the Grand Cayman Parrot.

Diet: Sea Grapes, Red Birch berries and the flowers, seeds and berries of many other native plants.

Reproduction: Cayman Islands parrots lay an average of three eggs per clutch in a tree cavity.

Range: The Grand Cayman Parrot is endemic to the island of Grand Cayman, while the Cayman Brac Parrot is endemic to Cayman Brac.

Habitat: Dry forest, mangroves and agricultural land.

Far left: *The Cayman Brac Parrot has the nickname Stealth Parrot because of its near-perfect camouflage.*

Left: *The Grand Cayman Parrot often feeds on the fruit of the Red Birch tree.*

Cultural presence

The Cayman Brac Parrot is also known as the Stealth Parrot because it is very quiet, whereas the Grand Cayman Parrot is often heard before it is seen because of the loud squawk it makes.

Cayman Islands parrots face many threats – both natural and unnatural. Natural threats include predation by hawks and owls, as well as introduced animals such as rats and cats; imported birds which can spread disease and compete for food and nest sites; heavy rain, which can flood nest cavities; and sometimes a shortage of food brought about by both droughts and storms.

Hurricane Paloma in 2008 caused a 48 per cent population decline in Cayman Brac; the population has taken three to four years to recover, but is back to previous numbers.

Among the unnatural threats facing parrots are deforestation, poaching and shooting. Deforestation for development purposes is resulting in the loss of traditional nest sites.

Poachers, who take birds for the pet trade, often cut down trees to reach parrots' nests. 'This illegal and thoughtless practice destroys a nesting hole that would have been used for many years and seriously impacts the parrot's ability to reproduce,' according to the national bird section of *Cayman Islands National Symbols: Flora and Fauna*. 'Captive birds often die and those that do survive may lead lonely and unhappy lives in cages.'

In the past it was common for local families to have pet parrots. Today, however, in the Cayman Islands it is illegal to keep the birds as pets.

Many fruit farmers in the Cayman Islands regard parrots as pests. However, one such farmer, Otto Watler, says that only around 59 kg (130 lb) of up to 11,340 kg (25,000 lb) of mangos grown by him annually are destroyed by parrots. 'This is less than one half of a per cent of my crop,' he is quoted as saying in *Cayman Islands National Symbols*. 'I can give that little bit back to nature so that my children and their grandchildren will have the parrot in their skies. Yes, it is frustrating to see the parrot wasting and ruining fruit, but we need to look at the overall actual losses. In my opinion, they are not large.'

The national bird section of *Cayman Islands National Symbols* adds that: 'Since the Cayman Islands Parrot lives only here, Caymanians are the only ones who can save it.'

One organisation having a positive impact on the conservation of the nature and wildlife of the islands is the National Trust for the Cayman Islands. This membership based charity has been working to preserve Cayman's natural heritage since 1987. As of 2013, the National Trust preserves 112 sq km (42 sq miles) of natural habitat useful for parrots on both Grand Cayman and Cayman Brac. These protected areas include the Mastic Reserve and Brac Parrot Reserve. In total the National Trust has been able to bring 5 per cent of the country's land area under protection, with protected areas on all three islands. The Trust can be contacted about guided tours or public access to these areas.

The Cayman Parrot appeared on this stamp of the Cayman Islands in 1962.

In 1996, the Cayman Parrot appeared on the 80 cents Cayman Islands' stamp.

CHILE

Andean Condor

Vultur gryphus

The Andean Condor is one of two animals depicted on Chile's coat of arms adopted in 1834, the other being a Huemul, or South Andean Deer. Both creatures are shown wearing the golden crown of the Chilean Navy.

The Andean Condor has also appeared on stamps, such as the 2008 Torres del Paine National Park issue, as well as on coins and banknotes.

WHERE TO SEE Good views of condors can be had in the Sierra de Ramon mountains near Santiago, according to the thisisChile website. This also recommends heading for the El Romeral and El Volcan peaks in the Cajon del Maipo mountains in summer to see these mighty raptors.

NEAR THREATENED

IUCN: The overall population is estimated to be 10,000 mature individuals and thought to be declining.

Size: Length 100–130 cm (39.4–51.2 in). Wingspan up to 3.2 m (10.5 ft). The world's biggest raptor and South America's largest flying bird.

Description: Adults of both sexes have a bare head, predominantly black plumage and a conspicuous white ruff. Males, which have yellow eyes, are also endowed with a large comb and neck wattle. Females lack the comb and wattle, and have red eyes.

Diet: Mainly carrion. Scavenges on the carcasses of mammals, including tapir, deer, rodents and domestic livestock. Can tear off and swallow more than 6.8 kg (15 lb) of meat at one sitting.

Reproduction: One egg is laid on a cliff ledge or in a small cave. The chick is nurtured by its parents for up to two years and takes five to eight years to reach maturity.

Range: Native to eight South American countries (Argentina, Bolivia, Chile, Colombia, Ecuador, Paraguay, Peru and Venezuela), and breeds in almost all of them, including Colombia.

Habitat: Usually found well away from people among the peaks and valleys of the High Andes. Capable of soaring up to 5,500 m (18,045 ft).

The Andean Condor is one of the largest birds in the world, with a wingspan of up to 3.2 m (10 ft) in length.

Cultural presence

Given its high profile, it is not surprising that the Andean Condor's name has been used in various ways in Chilean society. Chile's currency, for example, was changed from the peso to the condor in 1925. Confusingly, both continued to be used for many years, 10 pesos being worth 1 condor. The peso and condor were replaced by the escudo in 1960, and the escudo by the new peso in 1975.

The Eye of the Condor is the name of a major new annual event held in La Parva – 50 km (31 miles) from Santiago – to find the best videos and photographs of the ski resort, support the local arts scene and provide a forum for discussing the role played in society by action sports and photography. The inaugural Eye of the Condor took place in 2011.

CÓNDOR is the title of a weekly Chilean newspaper published in German, while Australia-based Condor Blanco Mines Ltd is a mineral exploration and mining development company with various copper and gold interests near the mining city of Copiapo in northern Chile.

While some people hope to see Condors on the wing in Chile merely as part of a general sightseeing holiday in the Andes, keen birdwatchers will almost always have these mighty raptors at the top of their tick list. Although holidaymakers might well come across Condors by chance, birdwatchers want to know exactly where to go to stand the best possible chance of observing these iconic birds of prey. Fortunately there are various tour companies which take Condor aficionados to the areas where these birds are most likely to seen. In fact, Condor watching is a niche ecotourism market in the Andes.

Cascada Expediciones (www.cascada.travel) says on its website that anyone looking for Condors should start in central Chile and head south. Horse riding in the Andes around Santiago, Chile's capital city, is "another good way" of seeing Condors "wheeling overhead," says the tour operator. It adds: "Whilst your horse takes care of getting you from A to B, you can focus on scanning the hills for the unmistakable condor silhouette."

Cascada Expediciones also says that three male and two female captive-bred and radio-tagged Condors were released in 2012 near Rio Los Cipreses National Reserve, just over 100 km to the south of Santiago.

The Andean Condor was depicted in the four-value, National Zoo anniversary sheet.

Plenty of guanacos means an "abundance of Condors" in Torres del Paine National Park in Patagonia. Cascada Expediciones says as many as 12 Condors have been counted circling the same piece of carrion. Patagonia Wildlife Safari tours are run in the park by Cascada Expediciones "to help you locate and identify these Andean icons."

Torrest del Paine – a very popular park with visitors to Chile - is one of the places visited by Far South Expeditions (www.fsexpeditions.com) on its 19-day Chilean Birding Adventure. Andean Condors and Black-chested Buzzard-eagles are two target species in this gem of a wilderness.

BBC News reported in August 2013 that at least 20 Condors had been found poisoned near Los Andes, some 80 km east of Santiago. Two of the birds had died and the rest were recovering at the time.

Cóndor is the name of a weekly German language newspaper in Chile.

CHINA

Red-crowned Crane

Grus japonensis

Despite the fact that nearly two-thirds of people voted for the Red-crowned Crane as China's national bird on a series of Chinese websites in 2004, this species has never been formally adopted as the country's official avian icon. Why? Because, it is reported, the scientific name for this species, *Grus japonensis*, means Japanese Crane. Behind the selection process, which started in 2003, was the State Forestry Administration and Chinese Wildlife Protection Agency.

Crane images are common throughout China. They have been found in Shang Dynasty (16th–18th century BC) tombs and on Zhou Dynasty (1100–221 BC) bronzeware. Red-crowned Cranes have also appeared on stamps.

WHERE TO SEE Established in the late 1970s, Zhalong National Nature Reserve, 26 km (16.2 miles) east of Qiqihar City in Heilongjiang Province, supports one of the biggest breeding populations of Red-crowned Cranes, according to the International Crane Foundation (ICF).

ENDANGERED

IUCN: Only 1,700 mature individuals left in the wild. Although the Japanese population is stable, the one on mainland Asia continues to decline as a result of habitat degradation and destruction.

Other threats include fires, human harassment, poisoning from pesticide-treated grain and power lines.

Size: Length *c.* 158 cm (62 in).

Description: Large, mainly pure-white bird with a black face and neck and a red crown.

Diet: Omnivorous species, eating insects, aquatic invertebrates, fish, amphibians and rodents, as well as reeds, grasses, berries, corn and waste grain.

Reproduction: Nests in reed, grass and sedge marshes. Usually two eggs are laid.

Range: Breeds mainly in south-east Russia, north-east China and eastern Hokkaido, Japan. Russian and Chinese populations winter chiefly in the Yellow River Delta and coastal areas of Jiangsu Province, China. Japanese Red-crowned cranes are non-migratory.

Habitat: Wetlands – predominantly marshy areas during the breeding season, but a wider range of wetland habitats in winter, including rivers, coastal saltmarshes and mudflats, cultivated fields and aquaculture ponds.

Red-crowned Cranes spend most of their time walking and resting, while about a third of their time is spent preening, foraging and breeding.

Cultural presence

Given that Red-crowned Cranes are traditionally associated in China with such virtuous qualities and conditions as peace, harmony, happiness, nobility, longevity and even immortality, it is not surprising that these graceful, elegant birds are commonly depicted in Chinese paintings, tapestries and other decorative arts. In fact, crane art is very popular and includes colourful, scroll-mounted brush paintings of cranes in various settings.

The Fairy Crane Mosque – Xian he or Fairy Crane is another name for the Red-crowned Crane – in Yangzhou, Guangdong Province, is one of four famous ancient mosques in China. Built in 1275 to serve Arab traders and rebuilt twice during the Ming Dynasty, the mosque is said to resemble a crane in shape. Fairy Cranes are referred to in traditional Chinese poems and drawings.

Beautifully crafted flutes made from the wing bones of the Red-crowned Crane were unearthed in the late 1990s at the early Neolithic (Stone Age) site of Jiahu in China's Henan Province. Tonal analysis of the six complete flutes, which were Carbon 14-dated at between 7,000 and 9,000 years old, revealed that the holes in them corresponded to a scale remarkably similar to the Western eight-note scale, which begins with do, re, mi.

Statues of cranes flank the emperor's throne in Beijing's Forbidden City.

Renowned Chinese-American artist Professor Hung Liu hand-painted 80 Red-crowned Cranes onto 64 panels of glass for a 49 m (160 ft) *Going Away, Coming Home* public art installation unveiled in 2006 at Oakland International Airport in California, USA – a collaborative venture between Liu and Derix Glasstudios of Germany. The imagery was derived from a Northern Song Dynasty (960–1127) silk painting, *Auspicious Cranes*.

The Red-crowned Crane and the pine tree are symbols of longevity and immortality in China.

***Below:** A bronze statue of a Red-crowned Crane stands in front of the Hall of Supreme Harmony, in Beijing, China.*

The Red-crowned Crane has appeared on a number of different-value Chinese stamps over the years.

COLOMBIA

Andean Condor

Vultur gryphus

The Andean Condor is perched on top of the shield forming Colombia's national coat of arms, introduced in 1834. Symbolizing freedom, the bird is depicted with its wings outspread and holding a laurel wreath in its bill.

WHERE TO SEE Historically widespread in Colombia, the Andean Condor declined catastrophically to only about 10 pairs in the entire country prior to a release programme.

Between 1989 and 1991, 22 captive-bred condors were released into three protected areas of the Andes in Colombia: Chingaza National Park, Purace National Park and Chiles Indian Reservation. 'I think we can now confidently say that the Andean Condor will flourish in Colombia,' said zoologist Alan Lieberman, former curator of birds at San Diego Zoo Global, who was heavily involved with the reintroduction, 'partly because it's culturally important. The death of a condor was once a cause for celebration. Nowadays it's a national tragedy.'

NEAR THREATENED

IUCN: The overall population is estimated to be 10,000 mature individuals and thought to be declining.

Size: Length 100–130 cm (39.4–51.2 in); wingspan up to 3.2 m (10.5 ft). The world's biggest raptor and South America's largest flying bird.

Description: Adults of both sexes have a bare head, predominantly black plumage and conspicuous white ruff. Males, which have yellow eyes, are also endowed with a large comb and neck wattle. Females lack the comb and wattle, and have red eyes.

Diet: Mainly carrion. Scavenges on the carcasses of mammals, including tapir, deer, rodents and domestic livestock. Can tear off and swallow more than 6.8 kg (15 lb) of meat at one sitting.

Reproduction: One egg is laid on a cliff ledge or in a small cave. The chick is nurtured by its parents for up to two years and takes five to eight years to reach maturity.

Range: Native to eight South American countries (Argentina, Bolivia, Chile, Colombia, Ecuador, Paraguay, Peru and Venezuela), and breeds in almost all of them, including Colombia.

Habitat: Usually found well away from people among the peaks and valleys of the High Andes. Capable of soaring up to 5,500 m (18,045 ft).

Cultural presence

When it comes to icons, few birds equal the sheer size or pulling power of the Andean Condor. Traditionally associated with liberty, health and power, the species has been adopted as a national bird by more countries – Bolivia, Chile, Colombia and Ecuador – than any other.

Also an important symbol in Argentina and Peru, the Andean Condor has long been featured in Andean mythology. The Incas, for example, believed that it was a messenger of the gods and responsible for the daily rising of the sun. It has been the subject of Andean art since 2500 BC.

All manner of beliefs and superstitions have grown up surrounding this species. Both revered and feared, the bones and organs of the Andean Condor are used in traditional medicine. It is thought that this bird's stomach can cure breast cancer and that its eyes, once roasted, improve vision!

In a report entitled 'The Andean Condor: A Field Study', the fieldwork for which was carried out in June 1968 to June 1970, author Jerry McGahan says that this species has 'often played a dramatic role in the cosmology of Andean peoples. In some instances, the bird's role described the culture's history and destiny. Long before the rise of the Inca civilisation, some South American people painted and wove beautiful representations of the bird and buried them with their dead.'

The Incas believed that the bird was involved in creating parts of the Andean world and helped maintain the daily rhythm of the Earth and sun. 'After the Spanish conquered the Incas, both victor and conquered may have viewed the Condor as a symbol for the fate of Andean natives.'

McGahan goes on to say that

Above: *The Colombian coat of arms features an Andean Condor with outstretched wings.*

Above: *Colombia's President Juan Manuel Santos during a press conference in 2011.*

Left: *The coat of arms bearing the Andean Condor appears on many official items such as the national flag.*

whereas most city dwellers in Andean countries have never seen a condor, many country folk of the sierra or serranos 'are familiar with the bird and believe it has magical powers that can help them in their everyday lives. Like the eagle in North America, the condor represents strength and courage, but in a manner more profound and explicit.'

Andean Condors face a variety of threats, including habitat loss, flying into power lines and persecution. Ironically, some people mistakenly believe that they attack livestock rather than merely feeding on animals that are already dead.

The species has been the subject of a major captive-breeding and release programme since 1989. Nearly 70 zoo-raised birds have been released at five locations in Colombia (Chingaza, Purace, Chiles, Nevados and Siscuni) as a result of a highly successful collaboration between the American Zoo Association's Andean Condor Species Survival Plan, 18 zoological institutions (these provided eggs, chicks and resources) and the Colombian Ministry of the Environment.

The latter, in turn, has worked closely with four non-governmental organisations (NGOs) to release and monitor Andean Condors and to educate the public. It was discovered in 2012 that 36 of the original 67 released condors were alive and well, and that at least 10 young birds had been raised at three of the five reintroduction sites.

Fundacion Neotropical, a Colombian NGO, uses a volunteer group of 24 'condor guards' to teach local people about the natural history and conservation importance of condors. In 2010 alone, 394 students attended 12 mini-conferences in schools, and four workshops were held for condor guards to train them in field techniques such as radio-tracking, and the use of binoculars and telescopes.

The President, Juan Manuel Santos, holds the Colombian flag with athletes at the XVI Pan American Games in Mexico.

***Above:** The Andean Condor is a popular choice of costume at the traditional Advent Parade in Mongul, Colombia.*

The Andean Condor was depicted in a two-value, endangered animals set of stamps in Colombia in 1992.

COSTA RICA

Clay-coloured Thrush

CLAY-COLOURED ROBIN

Turdus grayi

Known locally as the *Yiguirro*, the Clay-coloured Thrush was chosen by the government as Costa Rica's national bird in January 1977. It was apparently adopted for various reasons – the male's melodious and familiar song, the fact that the bird is very common and its importance in Costa Rican folklore.

WHERE TO SEE This is a familiar, common and widely distributed bird. It is often found near human habitation.

LEAST CONCERN

IUCN: Despite the fact that the number of mature individuals has yet to be established, BirdLife International says the overall population is 'very large' and 'appears to be increasing'.

Size: Length 23–26.5 cm (9–10.4 in).

Description: For a national avian icon, the Clay-coloured Thrush is a pretty nondescript bird, being essentially brown all over – darkish brown above and pale clay-brown below. Its bill is olive-yellow. Sexes are similar.

Diet: Fruit and various invertebrates, including earthworms, slugs and insects.

Reproduction: Usually three eggs are laid in a large, cup-shaped nest in a banana or palm tree, or some other tree.

Range: Extremely large. Native to Belize, Colombia, Costa Rica, El Salvador, Guatemala, Honduras, Mexico, Nicaragua, Panama and the USA. (Scarce Rio Grande resident.)

Habitat: Wide-ranging – occurs in a variety of habitats from open, wooded areas, coconut groves, coffee plantations and other cultivated land to field edges, pastures and gardens.

Cultural presence

It is said that the species' spring song is regarded by farmers as heralding the rains of the growing season. Monica Quesada, writing on the Costaricatourism website, likens the Clay-coloured Thrush to the people of Costa Rica – 'little, brownish, gentle creatures' who smile the way the bird sings, 'powerfully and beautifully'.

The Clay-coloured Thrush has appeared on at least two Costa Rican stamps – one issued in 1984 and the other in 2010.

This Clay-coloured Thrush stamp was issued in Costa Rica in 2010.

Cuban Trogon

Priotelus temnurus

LEAST CONCERN

IUCN: Although the number of mature individuals is unknown, this species is said to be common and widespread in Cuba. The population is thought to be stable.

Size: Length 23–25 cm (9–9.8 in).

Description: Nominate race has a violet-blue crown and nape, reddish bill, off-white throat and breast, red belly and under-tail coverts and dark green back. Sexes are alike. There are two subspecies. The Cuban Trogon subspecies found on Cuba is the nominate subspecies, *Priotelus temnurus temnurus* or *Trogon t. t.*

Diet: Mainly flowers, but also buds, fruits and insects. Hovers in flycatcher fashion while feeding.

Reproduction: Three to four eggs are laid, often in a hole made by a woodpecker. Other natural cavities are also used.

Range: Endemic to Cuba.

Habitat: Both damp and dry forests. Most common in mountainous areas. Also frequents shrubby areas and copses near water.

The Cuban Trogon is known locally as the *Tocororo*, *Tocoloro* or *Guatini*. Its plumage includes the three colours of Cuba's national flag – red, white and blue. It is said that the trogon is like the people of Cuba in that it loves its freedom and cannot live in captivity.

WHERE TO SEE Among the places where the Cuban Trogon has been seen by birdwatchers are Alejandro de Humboldt National Park, near Baracoa, and Topes de Collantes nature reserve park in the Escambray Mountains.

Cultural presence

The Cuban Trogon has been featured on a number of Cuban stamps over the years, the most recent one – at the time of writing – issued in 2011 has a value of 85 Cuban pesos. The species has also appeared on at least one Cuban coin.

The beautiful Cuban Trogon was one of nine birds illustrated in a set of Ramon de la Sagra bird stamps in Cuba in 1971.

DENMARK

Mute Swan

Cygnus olor

The Mute Swan or *Knopsvane* was voted the national bird of Denmark in 1984 by viewers of a TV programme called *Dus med Dyrene*. It received 123,336 votes out of a total of 233,635 cast. What is surprising, however, is that before the poll to choose a national bird, Denmark's avian icon was the Skylark *Alauda arvensis*. The Mute Swan's popularity is unsurprising, given that it is the subject of one of Hans Christian Andersen's best-known and most-loved stories, 'The Ugly Duckling'.

Swans of one sort or another have long been used in brand names and logos (for example on the Nordic swan eco-label of Denmark, Norway, Sweden, Finland and Iceland) and have also been featured in works of art and on commemorative objects such as the 1976 Copenhagen Mother's Day plate.

LEAST CONCERN

IUCN: It was estimated in 2006 that the global population was c. 600,000–610,000 individuals. Overall, Mute Swan numbers are increasing. There were 4,500 to 5,000 breeding pairs in Denmark in 2000 – a far cry from the three or four pairs left in the 1920s as a result of hunting. Since the 1980s the Danish population of Mute Swans has been pretty stable, and from 1970 to 1978 Denmark was home to the world's largest breeding colony, in Ringkobing Fjord.

Size: Length 125–160 cm (49.2–63 in); wingspan 2.4 m (7ft 10.5 in); weight 6.6–15 kg (14.5–33 lb). A big bird.

Description: Long-necked. Adults are white, except for the orange-red bill and black forehead knob. Black, stout legs and webbed feet. Sexes are similar, although males are larger and have a more prominent 'knob' than females (especially during the breeding season), and a more intensely coloured bill.

Diet: Mainly aquatic vegetation and grain, but also invertebrates and small amphibians. Feeds on the surface of water and by 'upending' to reach submerged food.

Reproduction: Generally lays five to seven eggs in a huge, mound-like nest among reeds and on streambanks and riverbanks.

Range: Extremely large. Native to some 60 countries, including Denmark.

Habitat: Virtually anywhere there is water – marshes, reedbeds, lagoons, rivers and streams, ponds, lakes, reservoirs and gravel pits.

A Mute Swan in flight on a sunny day in summer or a frosty day in winter is a sight to behold – a vision in white against a cerulean sky.

Left: *Cygnets often hitch a ride downriver. Typically they stay with their parents until they are six months old.*

Below left: *Mute Swans build large, mound-like nests of vegetation on riverbanks and in other aquatic areas. The grey and white cygnets take 120 to 150 days to fledge and aren't sexually mature until three years old.*

WHERE TO SEE Mute Swans are common, widely distributed birds and just as likely to be seen in parks and other man-made environments, such as reservoirs, gravel pits, fishing lakes, village ponds, large bodies of water used for water sports and sheltered areas of harbours and ports, as in purely natural aquatic habitats, including slow-moving rivers, streams, creeks, lagoons, freshwater marshes, estuaries and fjords.

Cultural presence

Danish-born Andersen's fairy tale 'The Ugly Duckling' is one of more than 160 tales he wrote and relates how a brown, unattractive cygnet grew up to be a beautiful, graceful swan. Danes are rightly proud of Andersen's literary legacy and no doubt voted for the Mute Swan because of the link between this species and their most famous author.

The Mute Swan has a high public profile within Denmark. It appeared on Danish stamps in 1935, 1986 and 2005 (a four-value Hans Christian Andersen set), and on 10-kroner Danish mint coins in 2005. The latter was part of a five-coin series marking the 200th anniversary of famous Danish children's author Hans Christian Andersen (1805–1875). On the reverse of the 10-kroner coin is an image of a swan looking at its own reflection in the lake at Bregentvedt Manor, where Andersen penned his 'Ugly Duckling' story.

On the morrow the wild ducks saw him. "And who may you be?" said they; "you certainly are an ugly fellow, but we don't mind that so long as you don't want to marry into our family." Poor ugly duckling! he was only too happy to be allowed to eat and drink and sleep in the rushes.

XVI.

'The Ugly Duckling' story by Hans Christian Andersen was made famous in modern times by a film and a Danny Kaye song of the same name.

DOMINICA

Imperial Amazon

IMPERIAL PARROT

Amazona imperialis

Also known as the *Sisserou* (the Pride of Dominica), the Imperial Amazon's image adorns not only the island's national flag, but also its coat of arms, public seal, the Mace of the House of Assembly and Dominica's Honours for Meritorious Service to the Country. The bird also appears on the crest of Dominica State College and on postage stamps and tourist souvenirs.

WHERE TO SEE The species is found in Morne Diablotin and Morne Trois Pitons National Parks and in all forest reserves on Dominica.

Over the years, a number of sick or injured wild Imperial Parrots have been cared for at the Parrot Conservation and Research Centre at the Botanical Gardens in Roseau. A pair of adult birds, which could not be returned to the wild for different reasons, made history in 2010 by producing the first captive-raised Imperial Parrot. Imperial Amazons are mainly found in mountain rainforest. It is endemic to Dominica.

● **ENDANGERED**

IUCN: It is estimated that there are 350 to 450 wild Imperial Amazons. The species teetered on the brink of extinction following a devastating hurricane in 1979. Remarkably, given the slow rate of reproduction, numbers have recovered to pre-hurricane levels, and may even exceed them.

Size: Length 45–50 cm (17.7–19.7 in). Biggest Amazon in terms of length.

Description: A strikingly distinctive bird with a dark purple head, neck and underparts, and a green back.

Diet: Fruits, seeds, flowers, shoots and young stems, mainly from rainforest trees, epiphytes and vines.

Reproduction: Tree nest cavities can be several metres deep, although some are quite shallow (that is, less than a metre deep). Most cavity entrances are more than 30 m (98.4 ft) from the ground. May produce only a single chick every other year.

Range: Endemic to Dominica.

Habitat: Found only in old-growth montane forests on the island.

The Imperial Amazon is found mainly in high altitude rainforests.

Cultural presence

The Imperial Parrot – known and revered since before Columbus's time – is a bird of which Dominicans are extremely proud and fiercely protective. 'Dominica, which is the biggest and most pristine of the Windward Islands, has long been a beacon for conservation,' said Dr Paul R. Reillo, Director of the Florida-based Rare Species Conservatory Foundation. 'More than one-third of Dominica is protected by national parks and reserves. Parrot conservation especially is more than just a gesture. It's part of the government and part of the fabric of everyday life.' Dr Reillo has worked closely for many years with Stephen Durand and his colleagues within Dominica's Forestry, Wildlife and Parks Division of the Ministry of Agriculture and the Environment, studying and conserving the Imperial Amazon. Habitat loss, hunting for food, trapping for the cage-bird trade and hurricanes have all taken their toll on the species, but the Imperial Parrot's future is now looking brighter.

Affectionately known as the Pride of Dominica, the Imperial Amazon enjoys a high profile in government circles, its image adorning the country's national flag, coat of arms and such like.

Throughout May, thousands of children in Dominica celebrate the richness of Dominica's bird life (the island is home not just to one, but two endemic parrot species, the other being the Red-necked Parrot *Amazona arausiaca*) through the environmental outreach programme activities of the annual Caribbean Endemic Bird Festival. This is a collaborative joint venture between the Forestry, Wildlife and Parks Division, Rare Species Conservatory Foundation, local sponsors and the Society for the Conservation and Study of Caribbean Birds. Experienced foresters accompany schoolchildren on birdwatching trips and give talks.

The 12th annual Caribbean Endemic Bird Festival was launched in 2013 and the festival is sponsored by the Society for the Conservation and Study of Caribbean Birds (SCSCB).

Above: *A multicoloured gem, the Imperial Amazon can live to be more than 70 years old in captivity.*

Right: *The Imperial Amazon was one of 15 bird species featured in a set of birds of Dominica stamps issued in 1987. The national bird graced the highest denomination stamp, which had a face value of $10.*

Sick and injured Imperial Amazons have been looked after at Dominica's Parrot Conservation and Research Centre.

DOMINICAN REPUBLIC

Palmchat

Dulus dominicus

The Palmchat is the official national bird of the Dominican Republic. It is known as *Cigua Palmera* to Spanish-speaking people.

WHERE TO SEE This is a very common, widely distributed species, which may well be Hispaniola's most common bird. It has adapted well to man-made environments.

LEAST CONCERN

IUCN: Precise numbers are unknown, but the Palmchat's population appears to be stable.

Size: Length 18–20 cm (7–7.9 in).

Description: Dark olive-brown upperparts, lighter, heavily streaked, creamy-buff underparts, red eyes and a stout ivory bill. Sexes are similar.

Diet: Mainly fruits, but also flowers and some invertebrates.

Reproduction: Two to seven eggs are laid in a massive and untidy communal nest with separate breeding chambers. Usually built in Royal Palms *Roystonea hispaniolana*, these large, domed structures, made of twigs, can measure 1–2 sq m (10.8–21.5 sq ft). Nests typically accommodate four to 10 pairs of birds, but sometimes many more.

Range: Found only on the West Indies island of Hispaniola (that is, the Dominican Republic and Haiti). A year-round resident.

Habitat: From Royal Palm savannahs to farmland, parks and gardens.

Cultural presence

The Palmchat is a familiar bird in the Dominican Republic – partly because it is noisy and associates with others of its kind in flocks or family groups, and partly because of its large, conspicuous nests. The species appeared on stamps in 1964, 1993, 1996 and 2008.

The Palmchat appeared on a 3c stamp in 1964 as part of a three-value set of Dominican Republic bird stamps.

Common and widespread in the Dominican Republic, the easy-to-see Palmchat is a gregarious bird, for it lives and nests communally.

Andean Condor

Vultur gryphus

The Andean Condor is the official national bird of Ecuador and a key feature of Ecuador's coat of arms. It is shown perched above Mount Chimborazo, the country's highest peak, with its wings open, symbolising courage and strength. The coat of arms, in turn, forms part of the national flag – in use since 1860.

In 1991 the Ecuadorian National Congress declared 7 July to be National Condor Day following a campaign by Quito-based Aves & Conservacion (Corporacion Ornitologica del Ecuador), BirdLife International's partner in Ecuador.

WHERE TO SEE One of the haunts of the mighty Andean Condor is the huge and aptly named Condor Bioreserve. Situated to the east of Quito, Ecuador's capital city, this 5.4-million-acre bioreserve incorporates seven protected areas: Sumaco Napo-Galeras, Cotopaxi and Llanganates National Parks, Cofán-Bermejo, Cayambe-Coca and Antisana Reserves and Pasochoa Wildlife Refuge, plus various watershed protection areas and private reserves. The bioreserve features everything from snow-capped volcanoes and high altitude páramos grasslands to rivers and rainforests.

NEAR THREATENED

IUCN: The overall population is estimated to be 10,000 mature individuals and thought to be declining. It is thought there are only around 50 wild condors in Ecuador.

Size: Length 100–130 cm (39.4–51.2 in); wingspan up to 3.2 m (10.5 ft). The world's biggest raptor and South America's largest flying bird.

Description: Adults of both sexes have a bare head, predominantly black plumage and a conspicuous white ruff. Males, which have yellow eyes, are also endowed with a large comb and neck wattle. Females lack the comb and wattle, and have red eyes.

Diet: Mainly carrion. Scavenges on the carcasses of mammals, including tapir, deer, rodents and domestic livestock. Can tear off and swallow more than 6.8 kg (15 lb) of meat at one sitting.

Reproduction: One egg is laid on a cliff ledge or in a small cave. The chick is nurtured by its parents for up to two years and takes five to eight years to reach maturity.

Range: Native to eight South American countries (Argentina, Bolivia, Chile, Colombia, Ecuador, Paraguay, Peru and Venezuela), and breeds in almost all of them.

Habitat: Usually found well away from people among the peaks and valleys of the High Andes. Capable of soaring up to 5,500 m (18,045 ft).

Cultural presence

As in other Andean countries, the condor name has been widely used in Ecuador. For example, there is the vast Condor Bioreserve, which covers an area of more than 21,000 sq km (8,100 sq miles) and boasts more than 760 bird species, including the Andean Condor, and El Condor National Park, which was named after the Condor mountain range within it. There is also the Condor Huasi

Andean Condors produce a single chick every other year.

project, one of whose aims is to rehabilitate and release captive Andean Condors, and the Condor Trust for Education, which promotes education and training in Ecuador.

There are condor ponchos, condor T-shirts and condor pendants (many of these incorporate a *chakana* – a cross widely used in Ecuador). A bronze sculpture of a condor adorns the base of a monument in Quito's Independence Square.

Listed locally as Critically Endangered, the Andean Condor population in Ecuador is declining because of persecution, lack of food and poisoning. The National Strategy for the Conservation of the Andean Condor in Ecuador includes research, monitoring, captive breeding, environmental education and community involvement. Since 2010, the US-based Peregrine Fund, which is a member of the Andean Condor Working Group, has provided scientific advice to the Ministry of Environment of Ecuador on condor conservation and, since mid-2011, has been conducting collaborative research on the movements and habitat use of Andean Condors.

In July 2013, for the first time, an Andean Condor was released into the wild carrying a satellite transmitter in an effort to glean invaluable data on movements and habitat requirements, foraging behaviour and roosting sites. Named Felipe, the bird has become an ambassador for raising awareness about condor conservation.

A bronze condor sculpture rises high above Independence Square in Quito.

The Andean Condor has appeared on Ecuadorian stamps and banknotes and is also emblazoned on the back of a rare and highly collectable 1-condor gold coin minted only in 1928.

Above: *A 1958 Ecuador stamp featuring the Andean Condor.*

Left: *Ecuador's coat of arms incorporates an image of the Andean Condor, representing courage and strength.*

EL SALVADOR

Turquoise-browed Motmot

Eumomota superciliosa

The Turquoise-browed Motmot is the official national bird of El Salvador. It has various local names, the one used in El Salvador being *Torogoz*.

WHERE TO SEE The Turquoise-browed Motmot is a common bird and can therefore potentially be seen in a variety of places from gardens to forests.

LEAST CONCERN

IUCN: The population could be as high as half a million mature individuals overall.

Size: Length 33–38 cm (13–15 in).

Description: A long, black-tipped, racket-like tail is this bird's most striking feature. It is an exotic-looking species whose main plumage colours in both sexes are turquoise, green and brown. Other diagnostic features include a black facial mask and large bill.

Diet: Various insects such as butterflies, bees and dragonflies; spiders, worms and other invertebrates; small snakes and lizards; and fruits.

Reproduction: Four eggs are usually laid in a long burrow, at the end of which is the nest chamber. The burrow is excavated in an earth bank or low cliff.

Range: Very large. Native to six Central American countries: Costa Rica, El Salvador, Guatemala, Honduras, Mexico and Nicaragua.

Habitat: From woods and forests to thickets, plantations and gardens.

Cultural presence

The country's national rugby union **team**, Rugby Union Equipo Nacional de El Salvador, is nicknamed *Los Torogoces*, which means the Turquoise-browed Motmots.

One El Salvador business, known simply as Torogoz, was named after the species by founder Oscar Panameno in 1977 – he first saw the bird in the countryside in his youth.

In 2007 the Turquoise-browed Motmot was named the fifth most poetic bird by Stephen Colbert of *The Colbert Report* (a US satirical TV show).

The Turquoise-browed Motmot has appeared on stamps in El Salvador, including a special national symbols issue in 2009.

Above: *Known in El Salvador as Torogoz, the Turquoise-browed Motmot has lent its name to at least one business in the country.*

Below: *A 20c El Salvador bird stamp – one of nine issued in 1963 – featuring the Turquoise-browed Motmot.*

ESTONIA

Barn Swallow

Hirundo rustica

Estonia adopted the Barn Swallow as its official national bird in the early 1960s following a campaign by ornithologists. The Award of the Soaring Swallow is bestowed by Wikipedia on people who are judged to have contributed outstanding articles about Estonia to the online encyclopaedia.

WHERE TO SEE Few birds are as familiar to Estonians as the chattering Barn Swallow which graces their towns, villages and countryside from spring to autumn. Swallows are as synonymous with the lazy days of high summer as blue skies and warm sunshine. In Europe, at least, few sounds and sights are as pleasing as swallows on the wing, catching insects in mid-flight, taking mud from pools with which to build their cup-shaped nests and making countless trips to and from their nests to satiate the voracious appetite of their offspring.

According to the National Symbols of Estonia website, the Barn Swallow is a 'characteristic guest of Estonian homes... its call can be heard from practically every eave or barn in the country.'

LEAST CONCERN

IUCN: There are estimated to be 190 million mature individuals globally. The Barn Swallow is the most familiar and widespread of all swallow species.

Size: Length 18 cm (7 in).

Description: Males of the nominate subspecies have blue-black upperparts and breast band, a dark red forehead and throat, off-white underparts and a deeply forked tail with streamers.

Diet: Mainly insects, from aphids, parasitic wasps, horseflies, hoverflies and blowflies in the breeding season to flying ants and termites in winter. Most are caught in flight. Eighty insect families have been recorded in this species' diet, according to volume 9 of the *Handbook of the Birds of the World*. Non-breeding birds in South Africa even eat some acacia tree seeds.

Reproduction: Builds a cup-like nest of mud and plant fibres, which it 'cements' to the interior or exterior of a man-made structure such as a house, barn or other building. Clutch size is usually four or five eggs.

Range: Huge – 51.7 million sq km (20 million sq miles). Found in many parts of the world. Mainly a long-distance migrant. British birds, for example, winter in South Africa.

Habitat: Open countryside, pastures, farmland and riparian areas, to villages, towns and even cities in some places. Barn Swallows are commonly seen on telegraph wires – often in large numbers in autumn prior to the autumn migration southwards. This species often roosts in extremely large numbers in reedbeds.

Above: *Young swallows have voracious appetites, which mean their parents are constantly flying back and forth to the nest with bills full of insects.*

Left: *Barn Swallows are a joy to behold on the wing, ducking and diving, twisting and turning as they catch insects.*

Cultural presence

The Barn Swallow has long occupied a special place in people's hearts and is widely referred to in literature, for this attractive little bird is symbolic of spring and a harbinger of summer. Literary references to it span thousands of years, and the Greek philosopher Aristotle cautioned that 'one swallow does not make a summer', while Shakespeare declared that 'true hope is swift and flies with swallow's wings'.

The Barn Swallow has appeared on some of the banknotes and coins of Estonia's former kroon currency, which was replaced by the euro on 1 January 2001. The Barn Swallow has also appeared on stamps.

It is the logo of the Estonian Chamber of Agriculture and Commerce's TEM, or Approved Estonian Taste label, which is given to quality Estonian food products whose principal raw ingredient is of Estonian origin.

Above: *The Estonian currency – the krooni, has featured the national bird on a 100 krooni note.*

Above right: *This stamp, issued by Estonia in 2011, featured the Barn Swallow as Bird of the Year.*

Source: *Estonia Post*

Eurasian Oystercatcher

Haematopus ostralegus

Locally known in the Faroe Islands as the *Tjaldur*, the Eurasian Oystercatcher is an extremely common species on the islands and is the official avian icon.

WHERE TO SEE The Eurasian Oystercatcher is a harbinger of spring, arriving in the Faroe Islands in March for the breeding season and leaving in September after the young birds have fledged. This species, which is distinctive in terms of both its plumage and its piping call, can be seen and heard all over the Faroe Islands.

LEAST CONCERN

IUCN: The global population is estimated to be around 1.1 to 1.2 million individual birds.

Size: Length 40–47.5 cm (15.7–18.7 in).

Description: Breeding adults have a black head, back, upper breast and tail, a white lower breast, belly, rump and wingbar, red eyes, a long, orange-red bill and pink legs. There are four subspecies. The Eurasian Oystercatcher subspecies found in the Faroe Islands is the nominate race, *Haematopus ostralegus ostralegus*.

Diet: Mussels, other bivalve organisms, crabs, earthworms, insect larvae and caterpillars.

Reproduction: Two to five eggs are laid in a shallow scrape in the ground.

Range: Huge. A very widely distributed species. Native to around 100 countries/ territories.

Habitat: Mainly saltmarshes, shingle or sandy beaches and coastal mudflats.

Cultural presence

The Eurasian Oystercatcher was spotlighted by seaman, shipbuilder and poet Nólsoyar Páll (1766–1808/9), a Faroese national hero, in his most famous work, *Fuglakvaeoi*, or Ballad of the Birds. Its role in this poem is thought to be the reason for its popularity and selection as avian icon.

The Eurasian Oystercatcher has been featured on stamps in the Faroe Islands on several occasions – in 1977, 1988 and 2002 (in the latter case as part of a four-value set featuring eggs and chicks). It also appeared on the reverse of the 5-kronur coin issued in 2011.

Above: *The Eurasian Oystercatcher, one of 11 oystercatcher species, lays it eggs in a lined or unlined scrape on the ground.*

Left: *This stamp, depicting the Eurasian Oystercatcher, was issued in the Faroe Islands in 1977. It was one of three species in of set of birds.*

FINLAND

Whooper Swan

Cygnus cygnus

The Whooper Swan is said to be have been a Finnish emblem for centuries, but it was not until 1981 that it was announced as Finland's national bird by the Finnish Council of Bird Protection. 'There was no actual voting, even if some sources mention a public vote, but no one has ever questioned the choice as it's hard to think of a better one,' said Jan Sodersved, Communications Officer of BirdLife Finland.

Despite being protected by law as early as 1934, there were only an estimated 15 breeding pairs of Whooper Swans in Finland in the early 1950s. By 1975, the breeding population had increased to around 150 to 200 breeding pairs. BirdLife Finland says the attitude towards this species was gradually changed – thanks, especially, to a couple of books written by wildlife author and veterinarian Yrjo Kokko.

The graceful Whooper Swan enjoys a high visual profile in Finland. Present in large numbers in many parts of the country during the breeding season, the image of this eye-catching species has been used in various ways. Swans of all types have long been synonymous with grace, beauty, love and fidelity, and are featured in many myths and legends.

LEAST CONCERN

IUCN: Large population of 180,000 mature individuals. Increasing or stable numbers in some areas, decreasing or unknown in others. Overall trend unknown. The species was close to extinction in Finland in the 1950s, but there are now at least 5,000 breeding pairs in the country.

Size: Length 140–165 cm (55.1–65 in).

Description: Large, long-necked white swan with a distinctive black-and-yellow bill. Very similar to the Trumpeter Swan *Cygnus buccinator* of North America. Sometimes confused with the smaller Bewick's Swan *Cygnus columbianus*.

Diet: Aquatic plants, grasses, sedges and horsetails when breeding. Cereal grain, potatoes, turnips, acorns and other vegetable matter are also eaten in winter.

Reproduction: Builds a large nest of plant material in or adjacent to a shallow freshwater pool or lake, slow-flowing river or sheltered coastal location. The same nest may be used for several years. Three to seven eggs are laid.

Range: Iceland and Scandinavia to north-east Siberia during the breeding season. Temperate Europe and Asia – parts of the UK to coastal China and Japan – in winter.

Habitat: Pools, lakes, marshes, rivers and similar sites in predominantly wooded areas of the sub-Arctic during the breeding season, and mainly agricultural areas, usually near the coast, in winter.

WHERE TO SEE The Whooper Swan is widespread in Finland during the breeding season. Large numbers of birds are seen on migration in spring (April) and in autumn (September–November) along the northern part of the west coast. BirdLife Finland says the best migration site in terms of Whooper numbers – 1,000 to 2,000 individuals – is Liminka Bay near Oulu where many other bird species can be seen too. There are also many other places throughout Finland where a few hundred Whooper Swans at a time can be seen during the migration periods.

Birdlife Finland says the sight of a Whooper Swan or a crane building its nest in spring 'is something that one can never forget'.

It adds that the country's popularity with ornithologists is greatly enhanced by the 'incomparable sights in late May and in September–October of the mass migration offshore and of waterbirds on their way to and from the Arctic'.

Top: *An adult Whooper Swan with cygnets, swimming. The latter fledge at around 87 days and reach sexual maturity when they are about four years old.*

Above: *The Whooper dines on aquatic plants during the breeding season but has a more eclectic diet in winter when it also consumes crops and other plant material.*

Left: *Like its relative, the more familiar Mute Swan, the Whooper also builds a large nest of plant matter. Unlike the Mute, however, it uses mainly mosses and lichens.*

Cultural presence

The 2002 Whooper Swan 0.50 euro Finnish stamp by Erik Bruun.

Source: *© Itella Posti, Finland*

According to Finland's national saga, *Kalevala*, anyone who kills a swan will also meet an untimely end. In *Kalevala*, a swan lives in the underworld realm of the dead. Famous Finnish composer Jean Sibelius based his *Lemminkainen Suite* on the *Kalevala*. Part of the suite, 'The Swan of Tuonela', paints a musical picture about this mythical, sacred swan of the underworld, and how Lemminkainen was asked to kill it but died before he could do so as a result of being shot with a poisoned arrow.

A stylised Whooper Swan is one of the many highly collectable birds immortalised in hand-blown glass by highly acclaimed Finnish designer Professor Oiva Toikka.

The oldest known Finnish Whooper Swan was an individual initially ringed on 10 May 1985 in Ranua, Lapland, Finland, when the bird was at least two calendar years old. The ring was recovered on 3 July 2009 – just over 24 years later – when the swan's body was found in Pedersore, Vaasa, Finland.

The Whooper Swan was shown in flight on a 0.50-euro stamp issued by Finland in 2002. It is central to the logo of the Nordic Eco-label or Swan (the official environmental labelling scheme of the five Nordic countries aimed at helping consumers to buy environmentally sound products). It appears on the 10-mark commemorative coins released in 1995 to mark Finland's entry to the European Union, the 1-euro coin (the obverse side shows two swans flying over a Finnish lake) and the Bank of Finland's 200th anniversary commemorative 100-euro gold coin, issued in 2011. The latter shows the body and one wing of the Whooper Swan on the obverse side and the other wing on the reverse. Two Whooper Swans in flight appear on Finnish 1-euro coin cufflinks.

A Finnish lake landscape featuring the Whooper Swan was once on the main page of the Google Internet search engine. Designed by Sophia Foster-Dimino of Google HQ, this 'Google doodle' marked Finland's independence day.

Commemorative 100-euro gold coins minted in 2011 in Finland featured the Whooper Swan. The official name of the coin is "Bank of Finland 200 Years".

Image: Mint of Finland

Finland's national bird has also appeared on the country's banknotes.

Gallic Rooster (Red Junglefowl)

Gallus gallus

The Gallic Rooster has long been used as an emblem or symbol by wide-ranging bodies and organisations in France. For example, it has formed an integral part of the Seal of the French Republic since 1848; is emblazoned on the flag and coat of arms of the mainly French-speaking Walloon or Wallonia Region of southern Belgium; has been used on French Rooster 20-franc gold coins (minted from 1899 to 1914); has appeared on stamps; was the official mascot for the 1998 FIFA World Cup (won by France); and forms the logos of France's national rugby league team and Le Coq Sportif (Athletic Rooster) sports equipment company.

IUCN: Not evaluated

Size: Length 65–75 cm (25.6–29.5 in).

Description: A rooster is another name for a male chicken and is a term more widely used in the USA than in the UK. In the UK such a bird is commonly referred to as a cockerel or cock. Roosters are unmistakeable, sporting pillar-box red wattles and comb.

Diet: Given the chance, chickens forage for and eat a wide range of food, including seeds, grasses and insects. Many poultry keepers buy specially formulated balanced diets, the precise type used varying according to the ages of their birds. Chickens kept in back gardens and on smallholdings are also often given fresh fruits and vegetable scraps, as well as treats like mealworms.

Reproduction: Hens lay a variable number of eggs, according to their breed. Some lay a few dozen eggs annually, others more than 200.

Range: Does not occur naturally in the wild.

Habitat: Being a domesticated bird, the Gallic Rooster is common and widespread in a variety of situations – from gardens and smallholdings to mixed farms and specialist poultry farms. All domesticated chickens are believed to be descended from the tropical Red Junglefowl.

Far left: *Cockerels will usually have started to crow by the time they are four months.*

Left: *Hens typically produce four to seven eggs per clutch.*

Cultural presence

France is unusual in having a domesticated bird rather than a wild one as its national avian icon. The Gallic Rooster has been an important national icon in France for centuries, adorning church weathervanes since the early Middle Ages and decorating flags during the French Revolution of 1789. *Gallus gallus*, the Latin scientific name for this bird, means both rooster and an inhabitant of Gaul. Given that roosters crow noisily at daybreak, it is thought that they originally symbolised vigilance.

The Gallic Rooster fell out of favour during Napoleonic times, the Emperor taking the view that the rooster was powerless and therefore an inappropriate national symbol for an empire of France's stature. Napoleon replaced the Gallic Rooster with an eagle. The Gallic Rooster may have been down, but it certainly was not out, for its fortunes as a national icon were restored in the 1830s when the Duke of Orleans signed an order authorising the use of the bird on the National Guard's flags and uniform buttons.

Officially, the Gallic Rooster is not France's national bird. Unofficially, however, it most certainly is, considering its cultural and historical significance and the continuing widespread use of its image.

Above: *French football fans wear the national colours with the Gallic Rooster to show support for their team.*

Above right: *The Gallic Rooster appeared on gold coins in the late 19th and early 20th centuries.*

Above left: *This French stamp, issued around 1965, depicts the Gallic Rooster.*

Right: *France has a long tradition of cockfighting. This sign hangs outside the entrance to a cockfighting arena in Saint-Amand-Les-Eaux in northern France.*

Above: *The logo of Le Coq Sportif sports company features a blue Gallic Rooster.*

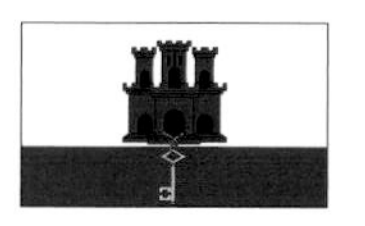

Barbary Partridge

Alectoris barbara

The Barbary Partridge is the official national bird of Gibraltar. It is also the emblem of the Gibraltar Ornithological and Natural History Society.

WHERE TO SEE A good area to look for the Barbary Partridge is the maquis and garrigue-covered upper part of the Rock, which is also a magnet for migrating birds in spring and autumn. Gibraltar is a noted migration hotspot: an estimated 250,000 birds of prey alone per season are funnelled across the narrow Strait of Gibraltar on their northwards or southwards journeys.

LEAST CONCERN

IUCN: The overall population has yet to be quantified. The European breeding population, which accounts for less than 5 per cent of the total, is thought to be between 7,500 and 20,000 pairs.

Size: Length 34–38 cm (13.3–15 in).

Description: Colourful, heavily marked, dumpy game bird. Bright red bill and largely grey upperparts, although the central forehead, crown and nape are a rich chestnut-brown. Black, white and rufous bands on flanks. Sexes are similar in colour. There are four subspecies. The Barbary Partridge subspecies found on Gibraltar is the nominate race, *Alectoris barbara*.

Diet: Mainly leaves, shoots, seeds and fruits. Also insects, especially ants.

Reproduction: Six to 20 eggs are laid in a lined depression on the ground.

Range: Principally North Africa.

Habitat: From rocky, arid hillsides, sand dunes and dry riverbeds to shrubland, woodland and citrus groves and other agricultural areas.

Cultural presence

The Barbary Partridge was introduced to Gibraltar from the Barbary Coast countries (the name is derived from the Berber people) of North Africa by British military personnel – probably for hunting purposes – in the 18th century. Gibraltar is the only part of mainland Europe where the Barbary Partridge breeds.

It has been depicted on various Gibraltar stamps – in 1960, 1991 and the 2010 set of bird definitives – as well as on some of the Rock's 1-penny coins.

Above: *A group of young Barbary Partridges is known as a covey.*

Left: *The Barbary Partridge appeared on this 59p Gibraltar stamp in 2010.*

GRENADA

Grenada Dove

Leptotila wellsi

Once known as the Pea or Well's Dove, the Grenada Dove was declared to be Grenada's national bird in 1991. The Grenada Dove is depicted on the island's coat of arms as a rather odd-looking blue-and-yellow bird opposite an armadillo.

WHERE TO SEE The species occurs in the following Important Bird Areas (IBAs): Beausejour/Grenville Vale, Mount Harman (where 43 per cent of the remaining birds are thought to live), Perseverance, Woodford and Woodlands. Mount Harman National Park was established in 1996 specifically to protect the Grenada Dove. The government of Grenada has committed itself to the creation of a second national park for this species at Beausejour, partly to offset the effects of a major tourism development in Mount Harman.

● **CRITICALLY ENDANGERED**

IUCN: Tiny population. Has declined as a result of habitat loss and fragmentation caused by hurricanes, fires and development (for example housing and roads). Predation by introduced animals such as rats, domestic cats and mongooses is also a problem. The maximum number of birds left in 2007 was estimated to be 136, according to BirdLife International.

Size: Length 28–31 cm (11–12.2 in).

Description: A 'twin-tone' bird – essentially brown above and white below with a pinkish-brown breast. White throat, a pale pink face and forehead, dark wings, a white-tipped tail, and pinkish legs and feet. Sexes are alike.

Diet: Unknown, but probably feeds on seeds. Forages only on the ground.

Reproduction: One nest, says volume 4 of the *Handbook of the Birds of the World*, was found on a palm frond 4 m (13 ft) above the ground. Lays two eggs.

Range: Occurs only on the island of Grenada in the Lesser Antilles. Sedentary.

Habitat: Mainly subtropical and tropical dry forest.

Cultural presence

The plight of the Grenada Dove, and what is being done to bring it back from the brink of extinction, has been spotlighted in schools and in ecotourism circles on Grenada. A 10-year conservation and recovery plan for the Grenada Dove was drafted in 2008.

It has a fairly high visual profile and has appeared on stamps.

This 75c stamp was one of four WWF Grenada Dove stamps issued in Grenada in 1995.

Resplendent Quetzal

Pharomachrus mocinno

Quetzal feathers, jade, precious stones, salt, obsidian (a natural glass of volcanic origin) and cacao beans were all used by the ancient Maya as currency. Such is the cultural and historic importance of the Resplendent Quetzal in Guatemala that it is not only the country's official national bird but also the name of its national currency. Introduced in 1925, each 'quetzal', or GTQ, consists of 100 cents – *centavos* in Spanish.

The distinctive image of a male Resplendent Quetzal, complete with flowing tail streamers, appears on all banknotes. The bird also forms part of Guatemala's national flag and coat of arms and has been depicted on a variety of stamps issued by the country since the late 19th century.

WHERE TO SEE One of the Resplendent Quetzal's haunts in Guatemala is the aptly named Quetzal Biotope or Mario Dary Rivera Nature Reserve between Guatemala City and Coban. Arguably the best-known place to see this species is the Monteverde Cloud Forest Preserve in Costa Rica.

NEAR THREATENED

IUCN: Up to 50,000 mature individuals. Deforestation is thought to be causing a moderately rapid decline.

Size: Length 36–40 cm (14.2–15.7 in).

Description: Extremely colourful and exotic-looking species. Males are truly spectacular birds, sporting long, emerald-green tail streamers that can add a further 65 cm (26 in) to their overall length.

Diet: Mainly fruits, but also insects, small frogs, lizards and snails.

Reproduction: Breeds March–August, but March–May in Guatemala. Lays one to two eggs in a deep, unlined cavity in a rotting tree trunk or stump. Territory 6–10 hectares (14.8–24.7 acres) in Guatemala.

Range: Mexico, Guatemala, El Salvador, Nicaragua, Costa Rica and western Panama.

Habitat: Undisturbed, humid, evergreen montane forests, cloud forests, heavily vegetated ravines and cliffs, park-like clearings, pastures and open, tree-scattered areas adjacent to forests.

The long-tailed Resplendent Quetzal is arguably one of the most beautiful and most exotic of all tropical birds.

A Resplendent Quetzal at the entrance to its tree cavity nest.

Cultural presence

The Resplendent Quetzal was venerated not only by the Maya, but also by the Aztecs. In fact, the name 'quetzal' derives from the Aztec word *quetzalli*, the original meaning of which was 'tail feather'. Quetzacoatl, a deity and cultural force for good worshipped by both cultures, was symbolized by the head of a snake

Even Guatemala's currency is named after the Resplendent Quetzal, the 'quetzal' having been introduced in 1925.

adorned with quetzal feathers. The Aztecs believed that Quetzacoatl would return in 1519 and rule over them as a god-king. When Spanish conquistador Hernan Cortes arrived in Veracruz that very year, the Aztecs assumed that he was Quetzacoatl in human form and promptly sent him a quetzal headdress.

Resplendent Quetzals must have been much more common in ancient times than they are today, for five Aztec provinces incorporating cloud forests were compelled to provide as many as 2,480 handfuls of mainly quetzal tail streamers as tributes. The *Handbook of the Birds of the World* states that if it is assumed that each handful contained 10 to 50 feathers, the Aztecs would have had to 'harvest' a staggering 6,200 to 31,000 quetzals annually. Even if the birds were released after their plumes had been removed (to kill a quetzal was a capital offence punishable by death), many are likely to have been badly injured in the capturing and plucking process.

For several decades, up to 800 Resplendent Quetzals were exported annually from Guatemala alone to meet the seemingly insatiable demand by Victorians for fashionable bird plumes. Nowadays Resplendent Quetzals are big business on the ecotourism front, and birdwatchers from all over the world travel to see these stunning birds in the wild.

One quirky and somewhat mysterious ancient link between the Maya and the Resplendent Quetzal was discovered in 1998 at the Temple of Kukulkan at Chichen Itza, Mexico, by US acoustics scientist David Lubman. He found that clapping in the open plaza of the Kukulkan pyramid generates anomalous echoes from temple staircases, the sonogram of which is very similar to that of a Resplendent Quetzal chirp. 'Only a single clap is necessary to evoke an audible chirped echo if the staircase is sufficiently sound reflective, such as the two restored staircases at Kukulkan pyramid.' A louder stimulus, like the sound generated by a synchronised clapping chorus, is

These two early definitives, issued in Guatemala in 1935, both showed the Resplendent Quetzal.

The Mayan character depicted in this statue is apparently wearing Resplendent Quetzal feathers as part of his costume.

required to produce echoes from the two unrestored staircases.

Lubman describes a situation in ancient times where a priest would be hosting a well-attended spring equinox ceremony at the Temple of Kukulkan. A manipulative priest could ask a question of the gods in front of his audience after which he would clap his hands. The sounds of the claps would be transformed in the quetzal chirps echoing around the temple. These chirp responses would be interpreted by the audience as a response from the gods which would reinforce the belief in the magical powers of the priests.

'The Maya faithful would probably recognise that the handclap sound emanated from a priest and the quetzal sound emanated from the temple. It would appear that the priest's aural inquiry was immediately rewarded with an aural answer from the gods relayed by the quetzal, messenger of the gods.'

Archaeolgists once dismissed this theory of the chirped echo 'as an artefact of reconstruction'. However, mounting evidence is now seen to support this as an intended design feature. Lubman says: 'Unfortunately, some archaeological scholars appear to be visually dominated and so ignore evidence of ancient uses of sound.'

The Order of the Quetzal, or Orden del Quetzal, is the highest honour that the government of Guatemala can bestow on its own citizens or foreigners. It is awarded in recognition of outstanding humanitarian, civic, artistic and scientific achievements.

One Order of the Quetzal recipient is Brown University (Providence, Rhode Island, USA) anthropologist, archaeologist and epigrapher Professor Stephen Houston, who received the Grand Cross rank of the award in 2011 in recognition of his 'extraordinary contributions' to the study of Maya culture.

Revered by the Maya and the Aztecs, the Resplendent Quetzal is highly regarded in modern Guatemala, for this spectacular species adorns the nation's coat of arms and flag, among other things.

Houston, who studies ancient Maya script to shed light on political and social systems in Mesoamerica, has worked on several major archaeological digs in Mayan cities, including the ancient city of Piedras Negras in Guatemala.

Houston said very few scholars had ever received the Order of the Quetzal and never before, to his knowledge, had any received the rank of Grand Cross.

"Guatemala is the country where I have spent close to 30 years a researcher of Maya civilisation," he remarked at the time. "It is the heart and focus of that civilisation from 1000 BC to the present. More than any place, Guatemala is where my heart lies."

Guatemala's national flag prominently features the Resplendent Quetzal sat upon a scroll with the date of Central American independence from Spain: 15 December 1821.

GUYANA

Hoatzin

Opisthocomus hoazin

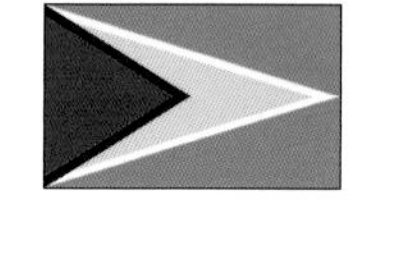

The Hoatzin was chosen as Guyana's avian icon more than 40 years ago because, according to the Guyana Tourism Authority, 'it was the most popular and easily identifiable bird species'.

The Hoatzin is depicted on the lower part of the shield of the country's coat of arms and is used as the official logo of the Guyana Tourism Authority. It has also appeared on stamps.

LEAST CONCERN

IUCN: The total number of Hoatzins is unknown, but it is thought that this species is common and its population is stable.

Size: Length 62–70 cm (24.4–27.6 in).

Description: The punk rocker of the bird world – a bizarre, multicoloured, primeval-looking avian oddity in a taxonomic group all of its own. Red eyes, bright blue, bare facial skin, a long rufous crest, a buff-white throat and breast, dark brown, white-streaked upperparts, chestnut thighs and wings, and a white-tipped long tail. Sexes are similar.

Diet: Strictly vegetarian, eating mainly leaves. The only known bird with a highly specialised, ruminant-like foregut fermentation digestive system. A slimy 'broth' of regurgitated vegetable matter is fed to chicks.

Reproduction: Breeding takes place during the rainy season. Two eggs are generally laid in an unlined platform nest of sticks and twigs in dense vegetation overhanging water. Young birds are raised not only by their highly social parents, but also by various 'helpers' – invariably older siblings.

Range: Very large – found in nine countries in northern South America, from Bolivia to Venezuela.

Habitat: Trees, bushes and vegetation next to lowland waterways.

Far left: *The Hoatzin is the punk rocker of the avian world – a prehistoric-looking bird that wouldn't seem out of place in Jurassic Park.*

Left: *The Hoatzin has a unique digestive system, regurgitating a revolting vegetable broth for its chicks.*

WHERE TO SEE A noted area to see the Hoatzin is the Berbice River and its Canje Creek tributary. Mahaica Creek, near Georgetown, Guyana's capital, has 'healthy populations' of Hoatzins, according to the Guyana Birding website.

Above: *The Hoatzin's distinctive crest is clearly visible in this photo of three resting birds.*

Left: *A poor flier, the Hoatzin is also known as the Stinkbird because of its leafy diet, which makes it smell of hay or fresh cow manure.*

Cultural presence

Many of the world's national birds are majestic, powerful, graceful or sweet-sounding. None of these qualities can be attributed to the Hoatzin. In fact, this prehistoric-looking, superficially pheasant-like bird spends much of its time sitting around digesting its food. Moreover, the Hoatzin climbs awkwardly, flies weakly and has a reputation for being rather smelly, the latter possibly having something to do with its unusual diet and digestive system.

According to the *Handbook of the Birds of the World*, few birds have a worse reputation for being, well, pongy. In Guyana the familiar Hoatzin is known not only as the Canje Pheasant, but also as the Stinkbird and Stinking Pheasant.

More than 100 international tour operators are now selling Guyana as a birdwatching destination. The Hoatzin is one of nearly 900 bird species in the country eagerly sought by amateurs and professionals alike.

Above: *The Hoatzin was one of four birds featured in a set of birds of Guyana stamps issued in 1990. Its image adorned the $60 stamp.*

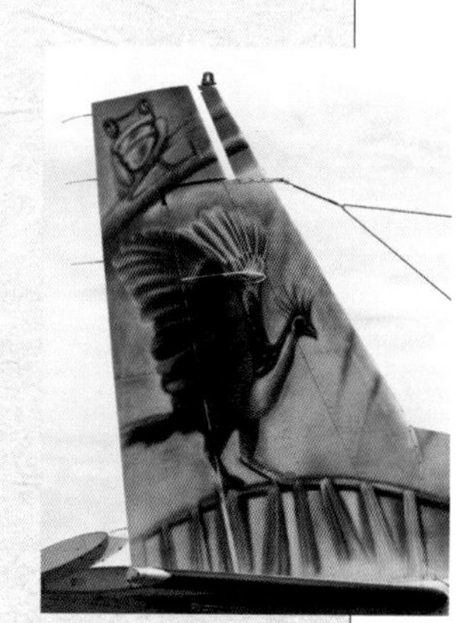

Right: *The tailfin of a Trans Guyana Airways aircraft features a painted Hoatzin.*

Hispaniolan Trogon

Priotelus roseigaster

NEAR THREATENED

IUCN: Ss a result of habitat loss and degradation. The global population has yet to be ascertained and is thought to be declining.

Size: Length 27–30 cm (10.6–11.8 in).

Description: A beautiful multicoloured bird with a yellow bill, orange eyes, a green crown and upperparts, a grey throat, breast and upper belly, and a red lower belly.

Diet: Chiefly insects, but also small lizards and fruits.

Reproduction: Two eggs are laid, often in an abandoned nest cavity chiselled out by a Hispaniolan Woodpecker *Melanerpes striatus*.

Range: Found only in Haiti and the Dominican Republic, the two countries making up Hispaniola. It is the only member of the trogon family on Hispaniola and one of only two trogon species in the entire Caribbean.

Habitat: Prefers forested mountains.

Despite being Haiti's official national bird, the Hispaniolan Trogon does not appear to have an especially high profile in the country. It forms part of the logo of the *Renmen* (the name means 'love' in Haiti) Project, the aim of which is to raise funds for UNICEF's Haiti Earthquake Children's Appeal. Limited edition prints of the Hispaniolan Trogon were produced by illustrator Sam Gilbey in support of the project.

WHERE TO SEE Aux Cornichons IBA on Massif de la Selle in Parc National La Visite, Bois Musicien IBA on the south-western edge of Macaya Biosphere Reserve in Massif de la Hotte and Aux Diablotins IBA on the northern border of Parc National La Visite.

Cultural presence

The Hispaniolan Trogon's habitat is increasingly coming under pressure, and BirdLife International has described Haiti as being 'one of the most environmentally degraded and densely populated countries in the world'. In fact, despite being national bird it is even sometimes hunted.

The Hispaniolan Trogon has appeared on stamps in Haiti.

Above: *The cavity-nesting Hispaniolan Trogon – a smorgasbord of colours – is one of only two trogons occurring in the Caribbean.*

Left: *Two bird stamps issued by Haiti in 1969 featured the Hispaniolan Trogon. One had a face value of 10c and the other (shown here) was 50c.*

Scarlet Macaw

Ara macao

Called *La Guara Roja* in Spanish and *Apu Pauni* by Moskitians, the Scarlet Macaw was chosen as Hondura's national avian icon on 28 June 1993 by the National Congress of Honduras to raise awareness of the country's rich bird life.

WHERE TO SEE With so few Scarlet Macaws left in the wild in Honduras, many bird lovers and tourists go to Macaw Mountain Bird Park and Nature Reserve in Copán Ruinas where rescued or donated birds can be seen at close quarters. It describes the macaws and other birds in the park as 'regional treasures', and says the park gives visitors 'a nearly complete reference of the parrots and toucans found in Honduras'.

LEAST CONCERN

IUCN: Precise numbers are unknown, but it is estimated that there are only 20,000 to 50,000 mature birds left in the wild. The population is declining overall as a result of habitat loss and poaching. There are two subspecies, with Scarlet Macaws in Honduras being of the *A. m. cyanoptera* subspecies. This is threatened but has not been officially designated as such by the IUCN. There could be as few as 200 or so Scarlet Macaws left in Honduras.

Size: Length 84–89 cm (33–35 in).

Description: A large and very colourful bird. Red, blue and yellow with a bare facial patch, prominent white bill and long, pointed tail. The Honduran subspecies differs from the nominate in that some or most of its yellow wing coverts have blue tips, and there is no green band separating the yellow and blue parts of its feathers.

Diet: Fruits, nuts, seeds, flowers and nectar.

Reproduction: Two to four eggs are laid in a large natural tree cavity.

Range: Very large. This species is native to 19 Central and South American countries, although the Central American subspecies is found only from southern Mexico to Nicaragua.

Habitat: Lowland rainforest and savannah gallery woodland. Found at an altitude of up to 1,100 m (3,608 ft) in Honduras. Roosts communally.

Far left: *Scarlet Macaws at a clay lick; they apparently consume the clay to obtain sodium.*

Left: *Multicoloured macaws are a spectacular sight in flight.*

Cultural presence

Macaws were revered by the Maya and commonly depicted in their art from around the 2nd century onwards. The best macaw sculptures are to be found at Parque Arqueólogico Ruinas de Copán, Honduras, a UNESCO-designated World Heritage Site, once described as the Athens of the New World. Among the exhibits at the government-managed Sculpture Museum of Copan are stone-carved macaw heads, a ball-court marker and a sculpture of a macaw in flight.

The Maya were captivated by macaws and other birds that lived around them. They were enchanted not only by the beauty of their plumage and the exquisite sounds they made, but also by their ability to fly. Feathers from birds such as macaws and quetzals were highly valued and traded as adornments for costumes.

The Maya traded extensively in Scarlet Macaws, bones from these birds having been found in association with the remains of Pueblo settlements as far north as Colorado, USA. It appears that the Maya bred Scarlet Macaws in Chihuahua, Mexico – at least 700 km (435 miles) from their native range. Some people have theorised that these birds were bred and traded so that they could be sacrificed at the end of the dry season in a bid to bring rain.

With sponsorship from BOSS Orange, the World Parrot Trust – an organisation involved with Scarlet Macaw research and conservation work for many years – has teamed up with the Macaw Mountain Bird Park and Nature Reserve, the Honduran Institute of Anthropology and History, and the Copan Association to return Scarlet Macaws to Parque Arqueológico Ruinas de Copán. A feeding and release facility was opened at the archaeological park in 2011. An associated educational programme has reached thousands of children at schools in the area. There is also a population of released Scarlet Macaws south of Tegucigalpa.

Dr LoraKim Joyner, Director of Florida-based Lafeber Conservation and Wildlife and of One Earth Conservation, and others are trying to save the Scarlet Macaw in La Moskitia, Honduras. Among those involved are Hector Portillo Reyes and Maria Eugenia Mondragon Hung of INCIBIO ICF (Forestry Service of

A close-up of a stone-carved Mayan Scarlet Macaw sculpture at the Mayan ruins of Copán, Honduras.

A researcher scales a large pine to measure a nest site that has been created to house Scarlet Macaws.

Honduras) and the indigenous villages of Rus, Rus and Mabita, La Moskitia. The work at present entails monitoring and protecting the Scarlet Macaw population and studying nest sites. A biological research station was due to be built in 2013. Born Free USA, which has contributed funds to the project and warmly welcomes donations, says a successful research station will result in not only a 'healthy population of Scarlet Macaws but also the preservation of rainforest habitat that many other wild creatures of many species and sizes call home'. It adds that Scarlet Macaws 'are under constant and sustained threat from deforestation and poaching'.

Tomas Manzanares, a local leader among indigenous people, has been shot four times while trying to protect communally owned land. Villagers, says Dr Joyner, 'fear to hunt and walk along through the pine savannahs and forests of their ancestors, but they will do so for the sake of their macaws'.

A number of Scarlet Macaw chicks confiscated from poachers have been raised and released into the wild.

The Scarlet Macaw has appeared on Honduran stamps and is the logo of Macaw Mountain Bird Park and Nature Reserve in the western highlands of Honduras.

Above: *A pair of Macaw chicks rescued from poachers are monitored before being released back into the wild.*

Above: *Dr Joyner and the team run a research station in La Moskitia, Honduras, to protect Scarlet Macaws.*

Left: *The local residents of La Moskitia help to protect the Scarlet Macaws, such as this juvenile.*

Right: *The Scarlet Macaw was one of four bird species illustrated on a four-value set of America stamps in 2001.*

Great Bustard

Otis tarda

The Great Bustard, White Stork and Saker Falcon all have relatively high profiles in Hungary for various reasons, although none of these birds appears to be either officially or unofficially the country's avian icon. Of these three species, the Great Bustard is a very important symbol of nature protection and draws many birdwatchers to Hungary. For this reason, it is profiled in depth in this section.

WHERE TO SEE There are 10 breeding populations of Great Bustards in Hungary, the most important in terms of numbers being at Kiskunsag, where there are around 550 birds – roughly one-third of the country's total. The second biggest population, with approximately 500 birds, is on the Devavanyai Plain.

VULNERABLE

IUCN: Numbers have declined significantly in Europe and beyond since the mid-1960s. Threats to this species include habitat loss and degradation, flying into power lines, destruction of nests by farm machinery, predation of eggs and young birds, and insufficient food for chicks. The global population is said to be 45,000 mature individuals.

Size: Length 105 cm (41.3 in) males, 75 cm (29.5 in) females; weight 5.8–18 kg (12.8–39.7 lb) males, 3.3–5.3 kg (7.3–11.7 lb) females. One of the world's heaviest flying birds. Females are smaller and much lighter than males.

Description: Males are brown and black above, white below, and have a reddish-brown breast band and a grey head and neck. Females are paler.

Diet: Plant material such as grasses and legumes and invertebrates.

Reproduction: Two or three eggs are laid on the ground – usually within wheat and alfalfa crops in Hungary.

Range: Found in many parts of the world, including the UK, Greece, Portugal, Afghanistan, Georgia, Russia and mainland China. The Great Bustard is native to nearly 40 countries and breeds in 15 European nations, including Hungary. Eastern populations are migratory, others partially so.

Habitat: Mainly flat or gently undulating, steppe-like grassy plains and agricultural land, but frequents cork-oak countryside and olive groves in south-west Iberia.

Left: *In the past, the Great Bustard often graced the tables of the European aristocracy.*

Left: *A male Great Bustard displays to a female during the mating season by ruffling its feathers and inflating a balloon-like structure in its neck. This mating display is called 'lekking'.*

Cultural presence

The Great Bustard has a high visual profile in Hungary. Known in Hungary as Tuzok, the Great Bustard is featured in several proverbs, including one stating: 'Better a sparrow today than a bustard tomorrow.'

The Great Bustard was once highly prized as a game bird, gracing the tables of European aristocracy. Nowadays, however, hunting is officially banned in all countries where this species occurs. The Great Bustard has been legally protected in Hungary since 1970.

This species forms part of the logo of the Hungarian Ornithological and Nature Conservation Society – BirdLife International's partner in Hungary.

An international action plan for the Western Palearctic population of Great Bustards, prepared by BirdLife International on behalf of the European Commission, calls for bird numbers to be restored to at least their 1979 levels. The document states that there were 1,582 breeding Great Bustards in Hungary in 2009, and that there had been a large decline over a 42-year period.

In a paper entitled 'The Great Bustard in Hungary', published in the April 2012 edition of *British Birds*, author Gergely Karoly Kovacs of the Hungarian Ornithological and Nature Conservation Society pointed out that Great Bustards are popular with visiting birders in his country and that low-impact tourism helps not only Great Bustard conservation but also the local economy.

Kovacs says that although the Great Bustard 'is not a national bird in Hungary, it is very important for nature protectors'. He adds that the White Stork is one of his country's best-known bird species and that the Saker Falcon is: 'treated as the "model" of a sacred mythological animal of Hungarian tribes 1,100–1,200 years ago. This bird was called Turul. In a myth, Turul makes love with a woman, Emese, while she sleeps. The baby born from this became the ancestor of the Hungarian nation. Later the outlook of Turul changed to an aggressive-looking, vulture-like bird of prey.'

The Great Bustard has appeared on a number of Hungarian stamps over the years, including a set of WWF bustard stamps in 1994 and a set of national parks stamps in 2001.

Above: *A 1971 World Hunting Exhibition stamp featuring the Great Bustard.*

Right: *The Great Bustard logo of MME – the Hungarian Ornithological and Nature Conservation Society.*

ICELAND

Gyr Falcon

Falco rusticolus

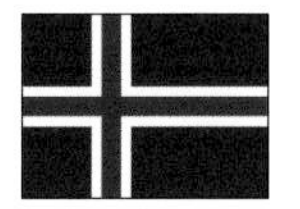

Long revered by falconers and once the subject of royal gifts, the Gyr Falcon was depicted on Iceland's coat of arms as a white bird on a blue background following a decree by the King of Denmark in 1903. This 'king's treasure' of a bird was considered to be a more appropriate symbol of the nation than the Icelandic cod hitherto featured.

Although the Gyr Falcon's heraldic reign was short-lived (a new coat of arms without it was introduced in 1919), the iconic raptor was chosen by decree in 1920 to appear on a special Icelandic royal flag. The species was also used in the 19th and early 20th centuries on the unofficial flag of Iceland's independence movement.

Discussions as to whether the Gyr Falcon should be reinstated on the coat of arms took place ahead of the creation of the Republic of Iceland in 1944, but it was unanimously decided not to make any changes.

The Gyr Falcon has been depicted on various Icelandic stamps,

LEAST CONCERN

IUCN: 110,000 mature individuals globally. The population is thought to be stable overall (there has been a 311 per cent increase in North America over 40 years).

Size: Length 48–60 cm (19–23.6 in). The world's biggest *Falco* species. Females are larger, heavier and sometimes darker than males.

Description: Bulky, broad-winged falcon. Very agile on the wing. Extremely variable in colour.

Diet: Mainly birds and mammals, which are taken on the ground or water at the end of a fast, low pursuit or seized in flight. The main food of Icelandic Gyr Falcons is the Rock Ptarmigan *Lagopus muta* game bird.

Reproduction: Up to one-third of all European Gyr Falcons breed in Iceland. The small and vulnerable Icelandic population is thought to number only around 400 breeding pairs in a good year. Usually three or four eggs are laid in a cliff-ledge scrape or in an old bird's nest – particularly one built by a Common Raven *Corvus corax*.

Range: Occurs in circumpolar Arctic and sub-Arctic regions from Alaska to Siberia. In some parts of its range the Gyr Falcon is migratory, while in other areas it is sedentary or partly migratory. Icelandic Gyr Falcons are sedentary, although juveniles travel all over the island and can be found in coastal areas in late winter.

Habitat: Cliffs, rocky coasts, offshore islands, craggy uplands, open country and even farmland, from sea level to around 1,500 m (4,920 ft).

Left: *The Gyr Falcon is a fierce, fast-flying predator that hunts mainly Rock Ptarmigans.*

Below left: *Gyr Falcon – the world's biggest falcon – pictured with a hapless pigeon it has just downed and killed.*

including one issued in 1930 to mark the 1,000th anniversary of Iceland's parliament, a set of four issued in 1992 in conjunction with WWF and a stamp/miniature sheet issued in 2004 to commemorate the centenary of the establishment of home rule in Iceland.

The Gyr Falcon has appeared on various Icelandic banknotes – either in its own right or as part of Iceland's old 'white bird' coat of arms.

WHERE TO SEE Lake Myvatn ('midge lake') in northern Iceland is a good place to look for Gyr Falcons in summer, as several pairs hunt in the area.

Cultural presence

Gyr Falcons have long been used in falconry and occupied a special place in the affections of people who kept, bred and/or flew these dashing birds. So highly regarded were Gyr Falcons in the Middle Ages that only a king was allowed to hunt with them. Today they are very much in demand and command high prices.

Gyr Falcons were certainly special to King Christian X of Denmark and Iceland, for not only did this early to mid 20th-century monarch decree that the bird should be depicted on a special Icelandic royal flag, but he also stated that it should lend its name to the Order of the Falcon which he established in 1921 to acknowledge meritorious conduct and deeds. There are five Order of the Falcon grades: Knight's Cross (the one most commonly awarded), Commander's Cross, Commander's Cross with Star, Grand Cross and Chain with Grand Cross Breast Star (reserved for heads of state). The President of Iceland, who is also Grand Master of the Order of the Falcon, presents crosses to Icelandic citizens twice a year – on 1 January and 17 June. Any man or woman considered sufficiently worthy may be nominated for an Order of the Falcon award. A white falcon on a blue background forms the centrepiece of each cross.

Above: *This is one of two Gyr Falcon stamps issued in Iceland in 1930.*

Left: *Thirty-four granite bird eggs, including a Gyr Falcon egg, run along the coast at Djupivogur, Merry Bay, in Iceland.*

Indian Peafowl

COMMON PEAFOWL / BLUE PEAFOWL

Pavo cristatus

Chosen as India's national bird in 1963 and protected by law, the Indian Peafowl has been revered on the subcontinent for thousands of years. It is known as *Mayura* in Sanskrit. The bird's image has been widely used as a logo and is often featured in Indian art, on everything from pottery, paintings and sculpture to metalware, jewellery, textiles and ivory.

WHERE TO SEE As a common to very common, widely distributed bird, the Indian Peafowl is not difficult to see, often frequenting cultivated areas near towns and villages.

LEAST CONCERN

IUCN: The global population has yet to be established, but it is thought that this species is common to locally very common. **Size:** Length 180–230 cm (70.9–90.5 in). Males are bigger than females. The gorgeous train of the cock bird accounts for 140–160 cm (55–63 in) of its overall length.

Description: The cock is by far the most colourful of the sexes, with a blue head and breast, white belly and striking, bronze-green train dotted with eye-like features called ocelli. A crest surmounts the head.

Diet: Omnivorous – feeds on grain, fruits and insects, as well as small mammals and reptiles.

Reproduction: Normally lays three to six eggs, which are hidden in scrub.

Range: Huge. Native to Bangladesh, Bhutan, India, Nepal, Pakistan and Sri Lanka. Introduced to various countries, including Australia, the UK and the USA.

Habitat: Frequents a wide range of terrain, including forest thickets near streams, secondary vegetation and cultivated areas.

Far left: *Peafowl hens and chicks are positively drab compared with the gaudy, much more colourful, cock birds.*

Left: *The long train consists of elongated upper-tail covert feathers. When fanned it is used to attract a peafowl hen.*

Cultural presence

According to the ancient Sanskrit epic the *Ramayana*, the god king Indra sought refuge under a peacock's wing after failing to defeat the many-armed, multi-headed Hindu demon Ravana. As an expression of gratitude, Indra endowed the peacock with 1,000 eyes – a reference, presumably, to the eye-like ocelli adorning the cock bird's tail. In another tale, Indra actually becomes a peacock.

Karttikeya, who is the main deity of the Yaudheyas and is always linked to war, weapons and hunting, is depicted riding a peacock on the reverse of gold coins issued by Kumara Gupta I (AD 415–450) and feeding a peacock on the obverse.

The peacock was associated not only with Karttikeya but also with many other Hindu deities and has been depicted on coins throughout India's long history. It is also featured in poetry, music and temple art.

The Peacock Throne of India, originally created for the 17th-century Mughal emperor Shah Jahan, was stolen from Delhi by the Persians in 1739. Said to have been one of the most magnificent thrones of all time, it featured the fanned tails of two peacocks – both of them gilded, enamelled and jewel encrusted.

Arjuna Nritham (the dance of Arjuna) is a ritual temple dance of Kerala in India. The costumes feature Peafowl covert feathers.

A symbol of pride, grace and beauty, the peacock is renowned for its ostentatious courtship dance, during which it raises and fans its tail in a bid to impress the much less attractive peahen.

'To peacock' was a term used in India in the mid-19th century to describe the practice of visiting ladies and gentlemen in the morning. If someone is said to be a 'peacock', it means that they are extremely proud or vain, while 'peacockery' is an adjective used to describe someone who is showy, conceited or puts on airs and graces.

This peacock decoration is inside the City Palace of Udaipur, India.

Fascinated by peacocks since childhood, Krishna Lal researched and wrote a book entitled *Peacock in Indian Art, Thought and Literature.* It was published in 2006 by Abhinav Publications of New Delhi.

The Indian Peafowl's iconic image has been used in a variety of ways and on all manner of objects down the ages. It has appeared, for example, on coins (both modern and ancient, the latter including Gupta Dynasty gold dinars and punch-marked silver coins of the Mauryan Empire), banknotes and stamps.

Peafowls appear throughout time and on a range of objects, such as this incense burner from the early 19th century.

State and territory emblems

Indian Roller

ANDHRA PRADESH
Indian Roller *Coracias benghalensis* Like all 12 roller species, this is a beautiful bird that adds a splash of colour to the farmland, plantations, palm groves and other habitats it frequents.

ARUNACHAL PRADESH
Great Hornbill *Buceros bicornis* Near Threatened, this hornbill is thought to be declining at a moderately rapid rate as a result of habitat loss/degradation and persecution by hunters and trappers.

ASSAM
White-winged Duck (also known as White-winged Wood Duck) *Cairina scutulata* is endangered.

BIHAR
House Sparrow *Passer domesticus* It was announced in 2013 that the House Sparrow, which is endangered in Bihar, would replace the Indian Roller *Coracias benghalensis* as state bird.

CHANDIGARH
Indian Grey Hornbill *Ocyceros birostris* A small, grey-and-white hornbill with an extremely large range.

CHHATTISGARH
Hill Myna *Gracula religiosa* One of Asia's most popular pets because of its ability to mimic human speech and other sounds. More than 170,000 wild-caught individuals have been exported from the countries in which the Hill Mynah occurs.

DELHI NATIONAL CAPITAL TERRITORY
House Sparrow *Passer domesticus* This species was declared to be Delhi's official bird by the government's chief minister in 2012 – partly, at least, to spotlight the fact that the once-common House Sparrow is declining in the territory.

GOA
Black-crested Bulbul *Pycnonotus melanicterus gularis* Although classified as being of Least Concern, this species is thought to be declining.

GUJARAT
Greater Flamingo *Phoenicopterus roseus* This is the biggest of all flamingo species.

HARYANA
Black Francolin *Francolinus francolinus* It has an extremely large range.

HIMACHAL PRADESH
Western Tragopan *Tragopan melanocephalus* Found only in high altitude forests in parts of India and Pakistan. Classified as Vulnerable because of its small and sparsely distributed population – an estimated 3,300 mature individuals.

Black-necked Crane

Asian Paradise-flycatcher

JAMMU & KASHMIR
Black-necked Crane *Grus nigricollis* A largely white-grey bird with a black head and neck and red crown patch. Classified as Vulnerable, the Black-necked Crane has an estimated global population of less than 11,000 individuals and is thought to have declined as a result of habitat loss and degradation.

JHARKHAND
Asian Koel *Eudynamys scolopaceus* This is a large long-tailed cuckoo. Males of the nominate race are bluish-black and glossy. The koel is featured widely in Indian poetry.

KARNATAKA
Indian Roller *Coracias benghalensis*

KERALA
Great Hornbill *Buceros bicornis*

LAKSHADWEEP
Sooty Tern *Sterna fuscata* A striking black-and-white bird with a strongly forked tail. The Sooty Tern is one of the commonest of all seabirds, its global population estimated to be more than 25 million pairs.

MADHYA PRADESH
Asian Paradise-flycatcher *Terpsiphone paradisi* There are many subspecies and plumage variations, but all male birds have a very long tail.

MAHARASHTRA
Green Imperial Pigeon *Ducula aenea*

MANIPUR
Hume's Pheasant *Syrmaticus humiae* This strikingly patterned pheasant is declining and classified as Near Threatened, its population being only 6,000 to 15,000 mature individuals.

MEGHALAYA
Hill Myna *Gracula religiosa*

MIZORAM
Hume's Pheasant *Syrmaticus humiae*

NAGALAND
Blyth's Tragopan *Tragopan blythii* Biggest and possibly the rarest of all tragopans. Classified as Vulnerable on account of the fact that its populations are believed to be small, fragmented and declining. Deforestation a major threat in north-east India.

ORISSA
Indian Roller *Coracias benghalensis*

Emerald Dove

Himalayan Monal

PUDUCHERRY (FORMERLY PONDICHERRY)
Asian Koel *Eudynamys scolopaceus* It was reported in 2007 that this long-tailed cuckoo had been adopted as Puducherry's territory bird.

PUNJAB
Eastern Goshawk *Accipiter gentilis* A large and powerful raptor, this goshawk species is known in Punjab as Baaz.

RAJASTHAN
Great Indian Bustard *Ardeotis nigriceps* This is a Critically Endangered species confined mainly to Rajasthan where, it is estimated, there are no more than 250 mature individuals. Brown and white with black head and wing markings, this bird has declined very rapidly.

SIKKIM
Blood Pheasant *Ithaginis cruentus* Found in forests, bamboo thickets and alpine scrub in the Himalayas and mountains of China, this species is declining throughout much of its range.

TAMIL NADU
Emerald Dove *Chalcophaps indica* A beautifully plumaged, largely forest-dwelling species that eats mainly seeds and fallen fruits.

TRIPURA
Green Imperial Pigeon *Ducula aenea* This species frequents a range of habitats, including wet evergreen and drier monsoon forests, cultivated areas, mangroves and Nipah Palm swamps.

UTTARAKHAND (FORMERLY UTTRANCHAL)
Himalayan Monal *Lophophorus impejanus*

UTTAR PRADESH
Sarus Crane *Grus antigone* This crane, which is believed to have declined rapidly, is found in northern and central areas of India, Nepal and Pakistan. Uttar Pradesh is this species' stronghold where, in 2008, it was estimated there were more than 8,000 individuals.

WEST BENGAL
White-throated Kingfisher *Halcyon smymensis* This medium-sized kingfisher eats mainly terrestrial insects.

White-throated Kingfisher

INDONESIA

Javan Hawk-eagle

Nisaetus bartelsi

● **ENDANGERED**

IUCN: There are only 600 to 900 mature individuals left in the wild. The population is believed to be declining moderately as a result of habitat degradation and destruction caused by growing numbers of people and the capture of birds for the pet market.

Size: Length 56–60 cm (22–23.6 in).

Description: A seriously impressive raptor with a prominent regal, white-tipped black crest that is often held upright. Male dark brown above; whitish, rufous-barred underparts; chestnut head sides and neck; long, grey-brown, banded tail.

Diet: Small mammals (for example tree shrews, bats and squirrels, rats and other rodents), as well as some birds, snakes and lizards.

Reproduction: One egg is laid in a large nest of sticks and leaves in a tree.

Range: Found only on the island of Java.

Habitat: Mainly tropical rainforests.

The national emblem of Indonesia, adopted in 1950, is a mythical golden eagle called Garuda, which is featured in both Hindu and Buddhist mythology. The design of this emblem is said to have been inspired by the *Elang Jawa*, or the Javan Hawk-eagle.

The Javan Hawk-eagle itself was declared Indonesia's National Rare/ Precious Animal – effectively its national bird – in 1993. Since then, its image has appeared on all manner of items, from stamps and banners to telephone directory covers.

WHERE TO SEE The Javan Hawk-eagle is widely distributed over much of Java, but has not been recorded in large parts of northern Java. Two areas where this species is found are Gunung Halimun Salak and Gunung Gede Pangrango National Parks in West Java. Gunung Halimun protects West Java's biggest montane rainforest, while Gunung Gede is Indonesia's most visited national park.

There were also two Javan Hawk-eagle pairs in 2011 in Grand Soerjo Forest Park, Batu City, East Java (down from about six pairs in 1997).

It is easy to see how the Javan Hawk-eagle was likely the inspiration for the design of Indonesia's national emblem, for male birds are predominantly brown and therefore not dissimilar to the mythical golden eagle, Garuda.

Cultural presence

The declaration of the Javan Hawk-eagle as Indonesia's National Rare/Precious Animal spotlighted this rare and endangered raptor in a way and to a level not previously seen, and made it the focal point of various conservation projects. What was not anticipated, however, was that greater public awareness would lead to a growing demand for Javan Hawk-eagles. It is reported that 30 to 40 individuals are sold at bird markets every year, although the actual number may well be much higher.

According to a scientific paper by Vincent Nijman, Chris Shepherd and S. van Balen, 'Declaration of the Javan Hawk-eagle as Indonesia's Rare Animal Impedes Conservation of the Species', published in *Oryx* in 2009, the Javan Hawk-eagle was one of the world's least-known birds of prey before the presidential declaration. The thinking behind the declaration made by President Suharto in 1993 was that it would be a catalyst for the protection of the species. Raising the profile of this species, however, also resulted in a 'great demand from zoos and malevolent collectors'.

Whether the Javan Hawk-eagles that appeared in zoos and bird parks following the National Rare/Precious Animal designation were already illegally owned by private individuals or traders, or whether the interest shown by zoos and fanciers 'created a larger demand, and hence an increase in captures, is unknown. If the latter supposition is correct, the declaration ... may have actually jeopardised the conservation of this species.'

Nijman *et al.* made several trade-curbing recommendations:

- Allow only a small number of authorised zoos and/or bird parks to display Javan Hawk-eagles.
- Require all captive birds to be micro-chipped and registered.
- Monitor bird markets regularly.
- Confiscate birds offered for sale and prosecute dealers.
- Educate the public that it is 'illegal and inappropriate to buy or keep protected species as pets'.

The Indonesian coat of arms features the mythical golden eagle Garuda, with a heraldic shield on its chest.

Above: *This Javan Hawk-eagle stamp was part of a six-value, environmental protection set issued in Indonesia in 1993.*

Left: *Indonesia's national airline is named after the country's mythical eagle of the same name. Garuda and the Javan Hawk-eagle are effectively one and the same bird.*

ProFauna, a non-profit organisation working for the protection of Indonesia's forests and wildlife, says it would like to mount a Javan Hawk-eagle campaign in the villages surrounding Grand Soerjo Forest Park in East Java, but had yet to receive any support at the time of writing. ProFauna always includes a picture of a Javan Hawk-eagle in its campaign material (such as posters and booklets) for protected species, which is displayed in schools, universities and other public places.

Another bird – a cassowary species – forms the centrepiece of the official emblem of the Indonesian province of Papua Barat on the island of New Guinea. It is unclear as to which of the three species of cassowary is depicted on the official emblem of the Indonesian province of Papua Barat (West Papua). All three species are found in New Guinea, though not necessarily in the same areas. It is certainly not the Dwarf Cassowary, which leaves the Southern Cassowary and Northern Cassowary. Although the representation on the emblem is stylised, it most closely resembles the northern Cassowary *Casuarius unappendiculatus*, which is found in Northern New Guinea. The critically endangered Bali Starling *Leucopsar rothschildi* became the official mascot of the Indonesian province of Bali in 1991.

The Rhinoceros Hornbill *Buceros rhinoceros* is the state bird of the Indonesian province of Sarawak. A symbol of strength and courage, the Rhinoceros Hornbill represents Singalang Burong – one of the most powerful Sea Dayak or Iban gods. Sarawak is known as Land of the Hornbills. The coat of arms of Sarawak depict what is described as a Great Hornbill, although the bird shown resembles a Rhinoceros Hornbill.

Above: *This Javan Hawk-eagle is being inspected for avian influenza and will be kept at the Yogyakarta Orangutan Centre for rehabilitation until it is ready for release into the wild.*

Left: *Large statues of Garuda, the mythical golden eagle, can be seen on buildings in Indonesia.*

Below: *In this Indonesian statue Vishnu, the supreme God of Hinduism, is depicted riding upon Garuda.*

Eurasian Hoopoe

Upupa epops

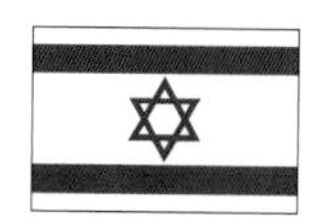

Called *Duchifat* in Israel, the Eurasian Hoopoe was chosen as the country's national bird in 2008 following a campaign initiated and led by Professor Yossi Yeshem of Tel Aviv University to draw attention to nature-conservation issues during Israel's golden jubilee year.

WHERE TO SEE Despite having distinctive plumage, the Hoopoe can blend in with its environment just like a tiger. It is often seen flying between trees or probing the ground, especially in grassy areas, with its long bill searching for invertebrates.

LEAST CONCERN

IUCN: Although there are an estimated five million mature individuals globally, this species is declining throughout its range as a result of habitat loss and excessive hunting.

Size: Length 26–32 cm (10.2–12.6 in).

Description: The Hoopoe is unlikely to be confused with any other bird, males of the nominate race having a long, decurved bill, an erectile crest, black-and-white wings and tail, and a sandy-pink head and body. Females are paler and slightly smaller. There are various subspecies, two of which are considered by some authorities to be separate species. There are nine subspecies of Hoopoe *Upupa epops*, the one found in Israel being the nominate race, *U. e. epops*. Often nicknamed the 'butterfly bird' because of the way it flies.

Diet: Mostly large insects and big, juicy grubs. Forages mainly on the ground, probing the soil with its long bill.

Reproduction: Four to eight eggs are laid in a natural cavity such as a hole in a tree, wall, cliff or old building, or even an unused rabbit warren.

Range: Huge. Found in many parts of the world, including China, Indonesia, Myanmar, Russia, the United Arab Emirates, the Mediterranean region, northern Europe and all over Africa.

Habitat: Open countryside, including dry, stony ground with low-growing vegetation, meadows, orchards, olive groves, cork-oak stands and even lawns.

The adult Hoopoe here is feeding its young an insect larva. Invertebrates make up the majority of the Hoopoe's diet.

Far left: *The female Eurasian Hoopoe lays approximately 8 to 12 eggs in a clutch. The nest will often be hidden in a cluster of rocks.*

Left: *The Eurasian Hoopoe is partial to invertebrates and insects, which it will often forage for on the ground, sticking its long beak as far as it will go into the soil.*

Cultural presence

The Eurasian Hoopoe was voted national bird in 2008. The selection process began in December 2007, when more than 1,000 bird enthusiasts attending a convention named the birds they considered to be most suitable as Israel's avian icon.

Members of the public were then invited to vote for their favourite among the 10 short-listed species (Barn Owl *Tyto alba*, European Goldfinch *Carduelis carduelis*, Graceful Warbler *Prinia gracilis*, Griffon Vulture *Gyps fulvus*, Eurasian Hoopoe, Lesser Kestrel *Falco naumanni*, Spur-winged Plover *Vanellus spinosus*, White-breasted Kingfisher *Halcyon smyrnensis*, Palestine Sunbird *Cinnyris oseus* and Yellow-vented Bulbul *Pycnonotus goiavier*). The chosen bird had to represent Israel's character, be a permanent resident of the country and feature in Jewish tradition. No fewer than 1.1 million people voted and the Eurasian Hoopoe was the winner. The campaign itself won two prestigious prizes as the best campaign of the year.

In 2009, a commemorative Israeli gold coin depicting the Hoopoe was minted.

Announcing the Eurasian Hoopoe's overwhelming victory, President Shimon Peres said that there was a great need for 'green scenery, fresh air and beautiful multicoloured birds', and lamented the fact that many once-common birds had become rare or had disappeared altogether from Israel. Peres disclosed that he had voted for the biblical vulture and was surprised that the dove had not been a finalist. An estimated 500 million birds pass through Israel in spring and autumn on migration.

Ironically, the Eurasian Hoopoe is not a kosher bird, for the Book of Leviticus in the Bible regards it and certain other winged wonders, like eagles, vultures and pelicans, as 'detestable, abhorrent' and not to be eaten. According to legend, the Hoopoe can cut through stone with its long bill, and when its bill breaks, it can cut instead with its folded crest. In fact, the name Duchifat means 'two beaks' in Aramaic.

The Hoopoe appeared on one of three bird stamps issued by Israel in 2010.

Ethiopian Jews called the Hoopoe the Moses Bird, believing that one day it would convey them to the holy city of Jerusalem.

It is said that a Hoopoe helped to lead King Solomon to the Queen of Sheba, and that a flock of the birds used their wings to protect the King from the sun.

The Eurasian Hoopoe appeared on gold and silver commemorative coins issued in 2009 to mark the 61st anniversary of the founding of the State of Israel. There were two versions of the silver coin and one of the gold.

The species was also featured on a 2010 stamp.

Red-billed Streamertail

WESTERN STREAMERTAIL / SWALLOW-TAILED HUMMINGBIRD

Trochilus polytmus

The Red-billed Streamertail is Jamaica's official national bird and as such enjoys high visibility in Jamaican society. Its image has appeared on stamps, from a 1956 pre-independence 6-pence definitive, to one of a number of 10-dollar BirdLife International stamps issued in 2004; on a 2-dollar banknote issued in 1969 and withdrawn in 1991; and on the reverse of a 25-cent coin withdrawn in 1997. It is also incorporated within Air Jamaica's new logo.

WHERE TO SEE Found throughout most of Jamaica, but especially common in the Blue Mountains. A noted hotspot is Rocklands Bird Sanctuary, about 5 km (3 miles) from Montego Bay, where Red-billed Streamertails perch on visitors' fingers to drink from feeder bottles. Much to the delight of Jamaican bird-lovers, Streamertails also frequent gardens (especially those with syrup feeders) and parks.

Like all hummingbirds, the Red-billed Streamertail is an iridescent avian jewel.

LEAST CONCERN

IUCN: Although the number of mature individuals is unknown, this species is described in the *Handbook of the Birds of the World* as a common resident.

Size: Length 22–30 cm (8.7–11.8 in) in males, of which 13–17 cm (6–6.7 in) is tail.

Description: Males are stunning, sporting an iridescent emerald-green body, black head, black-tipped red bill and forked black tail with very long streamers that make a distinctive humming or whining sound in flight.

Diet: Nectar and insects.

Reproduction: Breeds all-year round. Two eggs are laid in a cup-shaped nest lined with fine plant material and cobwebs. The nest is built on a thin twig 1–3 m (3.3–9.8 ft) above the ground.

Range: Endemic to Jamaica.

Habitat: All habitats at all altitudes, from man-made environments to dwarf or elfin forests. Most commonly found at an altitude of about 1,000 m (3,300 ft).

Left: *A Red-billed Streamertail hovers in mid-flight while 'sipping' nectar from a brightly coloured flower with its long, delicate tongue.*

Cultural presence

The Red-billed Streamertail is commonly known in Jamaica as the Doctor Bird. Why is not entirely clear, but one school of thought is that islanders felt that the adult male's black crest and long black tail resembled the top hats and long black cloaks once worn by doctors.

A Doctor Bird folk song widely mentioned on the Internet refers to 'Doctor Bud a cunny bud, hard bud fe dead' (sic), meaning that it is a clever, not easily killed bird.

According to Dr Rebecca Tortello, Jamaica's Tainos indigenous people called the Red-billed Streamertail the God Bird, believing it to be the reincarnation of dead souls.

In modern Jamaica, the name Doctor Bird is used in a variety of ways by all manner of businesses. For example, the Project Management Institute has a Jamaica Doctor Bird Chapter. Two power barges have even been named after this species (*Dr Bird I* and *Dr Bird II*). There is also the fruit-filled and cream cheese-frosted Doctor Bird Cake. The Doctor Bird, incidentally, was the mascot for the Eat Jamaican Campaign in 2011.

Jamaica's national bird was the inspiration for the ceramic Dr Bird Juicer, the centrepiece of which is shaped like the hummingbird, and is also mentioned in Ian Fleming's James Bond short story, *For Your Eyes Only*.

In her 'Delightful Dancing Dr Bird' poem, Karen Smith-Rose writes: 'Dancing on air as you go; a skip, a flip – a prize-winning show'.

Above: *A Streamertail appeared on this six-pence Jamaican definitive stamp issued in 1956.*

Below: *A Red-billed Streamertail adorns this $2 Jamaican banknote.*

Green Pheasant

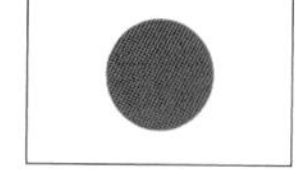

JAPANESE GREEN PHEASANT

Phasianus versicolor

Also known as the *Kiji*, the Green Pheasant was declared to be Japan's national bird in 1947 by the Ornithological Society of Japan after the Ministry of Education, Culture, Sports, Science and Technology decided that 4 April would be known as Bird Day.

WHERE TO SEE Not difficult to spot in a wide range of environments, including some places close to human habitation.

LEAST CONCERN

IUCN: Common and widely distributed on all of the main islands in the Japanese archipelago.

Size: Length 75–89 cm (29.5–35 in) males, 53–62 cm (20.1–24.4 in) females.

Description: Males are attractive, multicoloured birds, with a glossy black-purple head, throat and upper neck, bright red facial skin and blue-green underparts. Females are a warm sandy-brown with a variety of black markings. There are three subspecies of Green Pheasant *Phasianus versicolor* – *P. v. versicolor* (nominate race), *P. v. tanensis* and *P. v. robustipes*. All occur in Japan.

Diet: Thought to be a variety of seeds, fruits, berries and nuts.

Reproduction: Clutch of usually six to 12 eggs is laid in cover on the ground.

Range: Endemic to Japan. Introduced to parts of North America, including the main island of Hawaii, where it hybridises with the Ring-necked Pheasant *Phasianus colchicus*.

Habitat: Farmland, parks and other open areas.

Cultural presence

The Green Pheasant appeared on a set of five airmail stamps – each of a different value – in 1950. A pair of Green Pheasants adorned the reverse of the 10,000-yen banknote from 1984 to 1993.

The mascot of Fagiano Okayama – a football club in Okayama – is based on the Green Pheasant.

In some circles, however, the Red-crowned or Japanese Crane *Grus japonensis* – a species often featured on Japanese stamps – is considered to be Japan's national bird.

The 10,000 yen Japanese banknote issued in 1984 and 1993 featured a pair of Green Pheasants.

Sinai Rosefinch

PALE ROSEFINCH

Carpodacus synoicus

The Sinai Rosefinch is the national bird of Jordan, aptly the male's colouring matches the Jordanian sandstone landscape perfectly.

WHERE TO SEE This is a resident of the southern Rift Margins and Rum Desert. 'There can be no better place to see this bird than amongst the gorges and monument of Petra,' says Tareq Qaneer of Jordan's Royal Society for the Conservation of Nature.

LEAST CONCERN

IUCN: The Sinai Rosefinch's global population has yet to be quantified. It is said to be scarce to uncommon or locally common.

Size: Length 14.5–16 cm (5.7–6.3 in).

Description: Males of the nominate race are dapper birds with a rich, rosy-pink face and upper breast, light pink underparts and sandy-buff upperparts. Females are drab by comparison, usually lacking any pink. There are four subspecies of Sinai Rosefinch *Carpodacus synoicus*, the one occurring in south-western Jordan being the nominate race – *C. s. synoicus*.

Diet: Mainly seeds, buds, shoots and leaves.

Reproduction: Four or five eggs are laid in a large, cup-shaped nest tucked inside a rock or other crevice.

Range: Very large. Native to Afghanistan, mainland China, Egypt, Israel, Jordan, Palestine and Saudi Arabia.

Habitat: Arid rocky and scrubby terrain.

Above: *The world famous UNESCO World Heritage Site of Petra is one of the best places to see the Sinai Rosefinch.*

Above right: *The Sinai Rosefinch was one of eight species featured on a set of Jordanian bird stamps in 2009.*

Cultural presence

If ever a bird epitomised the environment in which it is found, it is surely the Sinai Rosefinch, for the male's beautiful pink plumage is a perfect colour match for the UNESCO World Heritage Site of Petra – the famous ancient city carved out of pink sandstone rock by the Natabaeans more than 2,000 years ago. It has appeared on four Jordanian stamps in 1987, 1997 and 2009.

LATVIA

White Wagtail

Motacilla alba

The White Wagtail was chosen as Latvia's national bird in 1960 by the International Bird Protection Council because of its popularity and admirable qualities.

Motacilla, part of this species' Latin or scientific name, is also the name of a company established by Latvijas Ornitologijas Biedriba (LOB) to promote and develop avian tourism in Latvia.

WHERE TO SEE The White Wagtail is a very common species in Latvia and can turn up in a wide range of habitats, including gardens, city parks, agricultural areas and the countryside generally, especially around human habitation.

LEAST CONCERN

IUCN: The global population is estimated to be anything from 50 to 500 million mature individuals.

Size: Length 16.5–18 cm (6.5–7 in).

Description: There are various subspecies, the one occurring in Latvia and other parts of mainland Europe being *M. a. alba*. The nominate race is grey, white and black, and a lighter colour overall than, for example, the *yarrellii* race – commonly known as the Pied Wagtail – which is found in Britain and Ireland.

Diet: All manner of terrestrial and aquatic invertebrates, including flies, dragonflies, beetles, spiders, ants and small snails. Readily takes household scraps.

Reproduction: Three to eight eggs are laid in a hair, feather or wool-lined cup of small twigs, grass stems and other vegetable matter.

Range: Huge. Found in many parts of the world.

Habitat: Everything from farmland and the wider countryside to gardens, parks and other town and city environments.

Cultural presence

The White Wagtail is the symbol of the Latvian Ornithological Society (Latvijas Ornitologijas Biedriba, or LOB), the biggest non-governmental nature-conservation body in Latvia.

Above: *The White Wagtail was also featured on a 2003 stamp in Latvia.*

Left: *The White Wagtail forms the centrepiece of the Latvian Ornithological Society's logo.*

LIBERIA

Common Bulbul

Pycnonotus barbatus

Although the Society for the Conservation of Nature in Liberia states that Liberia does not have a national bird, it would seem – if Internet references are anything to go by – that the Common Bulbul is at least commonly associated with Liberia and to some extent regarded as a symbol of the country. In Liberia it is better known as the Pepper Bird because of its apparent penchant for eating wild peppers.

LEAST CONCERN

IUCN: Although the Common Bulbul's overall population has yet to be determined, this species has been described as common to abundant. In fact, it is one of Africa's most common birds. Liberia's population alone is estimated to be at least 1.5 million pairs.

Size: Length 15–20 cm (5.9–7.9 in).

Description: A largely brown bird. Back, head, throat and upper breast grey-brown (head darker), belly whitish-grey. Dark bill, eyes and legs. Sexes are alike. There are 10 subspecies of Common Bulbul *Pycnonotus barbatus*. It is not clear from either the *Handbook of the Birds of the World* (HBW) or the BirdLife International species database which race or races are found in Liberia. HBW Vol 10 comments, in fact, that the whole group merits further study. There is extensive hybridisation.

Diet: Mainly fruits. Also invertebrates (for example termites, beetles, butterflies, dragonflies and grasshoppers), flower parts, nectar and small vertebrates such as geckos and lizards.

Reproduction: Two to five eggs are laid in a cup-like nest, generally 1–5 m (3.3–16.4 ft) above the ground.

Range: Very large. Native to nearly 50 countries/territories, including Liberia.

Habitat: Wide-ranging, including woods, bushes, forest margins, grasslands, riparian areas, plantations, gardens and orchards.

Left: *The Common Bulbul is very partial to fruit, including desert dates, figs, apricots and pomegranates. According to the* Handbook of the Birds of the World *(Vol 10), Common Bulbuls were recorded eating over 50 fruit species in one Zimbabwean study.*

Below: *The Common Bulbul is a very vocal bird. It makes a variety of sounds, including an alarm call, which has been described as resembling a machine gun.*

WHERE TO SEE The Common Bulbul can be seen almost anywhere in Liberia. An extremely familiar bird, it is commonly seen and heard in and around areas where people live. It takes food from garden feeders and also forages for insects on buildings.

Cultural presence

The Common Bulbul, or Pepper Bird, has a loud song and chattering, which has led to it being regarded as the alarm clock of West Africa.

According to legend, at dawn the Pepper Bird perches on Old Father Night's shoulder and urges him to return African children who have been lost in his safe and dreamy embrace to Father Day. Old Father Night, however, is reluctant to do this, not wanting children to face the burden and torment of Father Day. Instead, he holds them ever more tightly and pretends not to hear the increasingly loud shrieks of the Pepper Bird.

The not-for-profit Pepper Bird Foundation of Williamsburg, Virginia, USA, formed in 1987 to promote multi-cultural awareness among Americans, 'borrowed' its name from the Republic of Liberia which recognises 'the inimitable creature' that is the Pepper Bird 'as a national symbol'. The Foundation states that the Pepper Bird is noted for its ability to arouse and awaken. 'We send the Pepper Bird to arouse your interest and awaken in you a sense of history.'

The Land of the Pepper Bird: Liberia by Sidney De La Rue is a book about Liberia, while Pepper Bird Studios makes micro-financed feature films about the country.

The Common Bulbul appeared on Liberian stamps in 1953, 1977 and 1979 and was part of the logo of the now-defunct Air Liberia. A dove of indeterminate species, rather than a Common Bulbul, is featured on Liberia's flag.

The Common Bulbul is also the icon for the newsletter of the Society for the Conservation of Nature in Liberia (SCNL).

The Common Bulbul was one of six bird species featured on a set of Liberian stamps in 1953. The Bulbul and two others were depicted on triangular stamps.

LITHUANIA

White Stork

Ciconia ciconia

It has been Lithuania's national bird since 1973, with between 12,500 and 13,000 pairs nesting in Lithuania every year – the highest density of White Storks in any European country.

The annual return of storks to their favoured nest sites on buildings and electricity-supply poles in the countryside is eagerly anticipated and warmly welcomed, with 25 March being celebrated as national Stork Day. Enthusiasts carry White Stork cut-outs and don stork costumes on national Stork Day.

WHERE TO SEE White Storks are best seen during the breeding season when they are sitting on eggs and feeding young in their large and extremely conspicuous nests, or feeding in agricultural areas.

LEAST CONCERN

IUCN: There are 500,000 to 520,000 mature individuals globally. Numbers are increasing overall, although some populations are decreasing or stable.

Size: Length 110–115 cm (43.3–45.3 in). Very large bird.

Description: Long neck, bill and legs. Adults have a black and white body, and a bright red bill and legs. Unmistakeable, especially in flight.

Diet: Opportunistic. Takes a wide range of prey, including large insects, earthworms, small mammals, amphibians, reptiles and fish.

Reproduction: An average of four eggs are laid in a huge stick nest built either in a tree or on a man-made structure such as a roof, chimney, pylon, telegraph pole or electricity-supply pole (especially the latter). Following a decline in White Stork populations in some parts of western and southern Europe in recent decades, conservation efforts have included the building of man-made artificial nesting platforms in some countries.

Range: Extensive. Breeds mainly in Europe and winters mainly in Africa. European storks migrate south in late August in vast flocks that take advantage of thermal updraughts, usually arriving in Africa by early October.

Habitat: Open lowland. Tends to prefer wetter areas in its breeding territory and drier terrain in its wintering quarters. Occurs everywhere from damp or wet pastures, pools, marshy areas and arable fields to grasslands and savannah plains. Sometimes nests in towns and cities.

Cultural presence

The White Stork has long been close to the hearts of the Lithuanian people and is firmly embedded in the country's traditions and culture.

There are various stork sayings, beliefs and customs in Lithuania. For example, storks are supposed to bring babies or leave children in cabbages; people are honest wherever a stork's nest is found; there will be a good harvest if a stork nests in a harrow placed on a barn roof; a good year can be expected if a person sees a stork flying towards them; and there will be happiness in or near any homestead in which a stork settles. At one time country folk would hoist a cartwheel into a tree for the birds to use as a nest site, hammering copper coins into it so that storks could warm their feet on the metal on cold spring days.

The White Stork is the logo of the Lithuanian Ornithological Society (LOD). The LOD launched a four-year 'Conservation of the White Stork in Lithuania' project in 2009, the aim of which was to ensure 'long-term effective protection and favourable conservation status' for this species. Specific objectives included:

- Preparing a White Stork action plan.
- Conducting a detailed inventory of stork nests in Lithuania and creating a database of them.
- Erecting at least 1,760 nest platforms on overhead electricity poles to replace existing nests.
- Erecting at least 500 nest platforms on the roofs of buildings to replace problematic nests.
- Evaluating White Stork protection and identifying the key areas for this species.
- Increasing public awareness of White Stork biology, ecology and conservation.

The LOD declared 2010 to be the Year of the White Stork to raise the profile of and focus attention on the country's national bird and, in particular, to make people aware of the fact that stork habitat could be lost as a result of intensive agriculture and rapid economic growth. Around 1,000 people visited the Year of the White Stork festival in Vilnius, the Lithuanian capital. More than 300 Vilnius residents were 'ringed' and photographed inside the country's biggest artificial stork nest, measuring 20 sq m (215 sq ft), built in V. Kudirka Square.

Left: *This 2013 stamp depicts the White Stork as Lithuania's national bird.*

Above: *The Lithuanian Ornithological Society (LOD) has adopted the White Stork as its mascot.*

Below: *Bird-lovers and villagers show their support for their national bird at the Year of the White Stork festival in Vilnius.*

Despite being a popular and high-profile bird, the stork's image does not appear on Lithuania's flag. It was, however, depicted on aircraft of the now-defunct Air Lithuania regional airline and forms part of the logo of the Union of the Peasants and Greenroute. The White Stork has appeared on many public materials including stamps, posters and leaflets.

Goldcrest

Regulus regulus

The Goldcrest is the unofficial avian emblem of Luxembourg.

WHERE TO SEE A common bird likely to be heard and/or seen at any time of the year, from spruce and fir-dominated boreal forests during the spring and summer to a wider range of environments, including mixed and deciduous woodlands, parks and gardens, in winter.

Cultural presence

Given its small size and unobtrusive appearance, the Goldcrest seems a strange choice for a national bird. In fact, although the Goldcrest is Luxembourg's unofficial avian icon, it appears to have a low public profile within the country. In one respect, however, the Goldcrest is a singularly appropriate national bird for the Grand Duchy of Luxembourg. For, as Letzebueger Natura Vulleschutzliga (LNVL), BirdLife's Luxembourg partner, says: 'It was chosen because it's the smallest European bird and Luxembourg is Europe's smallest country.'

The name of LNVL's magazine was changed in 1954 to *Regulus* after the species' latin name, which means 'little king'.

The Goldcrest is generally a species that has to be actively sought out; it is quite difficult to come across accidentally. Anyone who is familiar with its high-pitched song will know that this beautiful little bird is actually more common and widespread than it might otherwise appear to be.

LEAST CONCERN

IUCN: The global population is thought to be 80 to 200 million mature individuals. It was estimated in 2004 that there were between 10,000 and 20,000 breeding pairs – equivalent to 30,000 to 60,000 individuals – in Europe.

Size: Length 9 cm (3.5 in). The Goldcrest is Europe's smallest bird.

Description: Diminutive songbird, of which there are 14 subspecies. The nominate race, *R. r. regulus*, is the one found in Europe and West Siberia. Males of this race have olive-green upperparts, greyish-white underparts, a bright yellow, central crown stripe sandwiched between a pair of black lateral stripes, two white wingbars, dark eyes and a needle-like bill. Females have a paler crown.

Diet: Small arthropods such as springtails, spiders, aphids and moths.

Reproduction: Six to 13 eggs are laid in a beautifully made, lichen-coated nest containing up to 2,500 small feathers.

Range: Huge. Native to more than 60 countries.

Habitat: Predominantly coniferous forests and woods in Europe during the breeding season, but a range of habitats at other times of the year (for example deciduous and mixed forests, and woods, shrubs, parks and gardens).

A stamp from 1970 with a Goldcrest to mark the 50th anniversary of the LPO (Protection des Oiseaux) bird protection organisation
Source: *Post Luxembourg*

African Fish-eagle

Haliaeetus vocifer

The African Fish-eagle appears on Malawi's coat of arms and is the national bird of Malawi. On the shield of the latter is a pair of wave-like blue lines separated by white bands. The bird, lines and bands together are symbolic of Lake Malawi – one of the biggest bodies of fresh water in Africa and one of the deepest lakes in the world. The African Fish-eagle is also used on the Malawi Police flag.

LEAST CONCERN

IUCN: Very large population of 300,000 mature individuals, which appears to be stable.

Size: Length 63–75 cm (24.8–29.5 in). Medium-sized fishing eagle.

Description: Adults are a striking combination of white, chestnut-brown and black. Long, broad wings, a short, rounded tail and a large head with a prominent bill.

Diet: Mainly live fish, which it swoops down on and grabs from the water in its talons. Also takes nestlings and even some adult birds. Piratical, harrying other birds such as storks, herons and kingfishers into dropping their prey.

Reproduction: Two eggs are usually laid in a large stick nest in a tree near water.

Range: Found over much of sub-Saharan Africa to the most southerly parts of South Africa. Locally common to uncommon or rare.

Habitat: Wide-ranging. Lakes, rivers, floodplains and stocked dams inland. Estuaries, creeks and mangrove lagoons in coastal areas.

Right: *Poised, ready for action, an African Fish-eagle comes in for the kill, its talons outstretched.*

Far right: *Like Ospreys, the African Fish-eagle is extremely adept at catching fish – flying low over the water and then plucking its hapless victim from just below the surface in its huge talons.*

Various Malawi stamps over the years have depicted the African Fish-eagle. In 1983, for example, a strip of five stamps of the same value showed this species in a variety of poses. The African Fish-eagle has also appeared on at least one kwacha coin.

WHERE TO SEE The best place to see the African Fish-eagle is undoubtedly Lake Malawi.

Above: *The beauty and strength of the African Fish-eagle is superbly captured in this image: big, powerful wings; striking, largely brown and white plumage; and formidable talons and bill.*

Left: *Lake Malawi, located between Malawi, Tanzania and Mozambique, is a favourite with African Fish-eagles. These birds are indigenous to sub-Saharan Africa.*

Cultural presence

The African Fish-eagle is a common bird not only in Malawi, but also in many other parts of Africa, so is a familiar sight to local people as they go about their everyday business.

Fishermen are more likely than most to encounter the species regularly, especially on and around Lake Malawi – a vast, wide inland 'sea' that is 587 km (365 miles) long and 84 km (52 miles) wide. According to Malawi Tourism, there are more African Fish-eagles at Lake Malawi than anywhere else in the world. They can often be seen plunging, talons-first, into the waters of the lake to snatch one of the countless fish that make their home in Africa's third biggest lake. They can also be heard calling from lakeside trees or as they fly low over the water.

Lake Malawi is renowned for its piscine biodiversity, all but five of its *mbuna* – African for rockfish – occurring nowhere else in the world. *Mbunas* are colourful, aggressive, freshwater fish popular with many aquarists. Overall, it is thought that there are around 1,000 different fish species in Lake Malawi – possibly more than in any other lake in the world.

Lake Malawi, where African Fish-eagles can be seen, is home to around 1,000 fish species.

MALTA

Blue Rock-thrush

Monticola solitarius

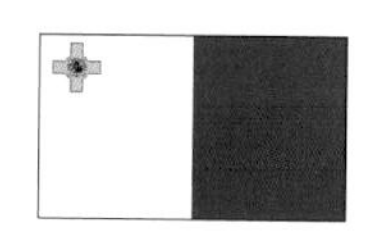

Known locally as the *Merill*, the Blue Rock-thrush was chosen as Malta's national bird by the Maltese authorities at the request of the Malta Ornithological Society (now BirdLife Malta). It was formerly the emblem of the Malta Ornithological Society.

LEAST CONCERN

IUCN: The number of breeding pairs in Europe was estimated by BirdLife International in 2004 to be between 120,000 and 260,000. The European population accounts for 25 to 49 per cent of all Blue Rock-thrushes, putting the global population roughly somewhere between 735,000 and 3,120,000 individuals.

Size: Length 20–23 cm (7.9–9 in).

Description: Essentially a dark blue bird. The female is variable and usually much duller than the male. There are four subspecies, the nominate *M. s. solitarius* being the one occurring in Malta

Diet: Mainly insects (from grasshoppers and locusts, to beetles, ants and flies), but also worms, snails, spiders, small lizards, snakes, frogs, toads and mice, as well as fruits including berries, and seeds.

Reproduction: Three to six eggs are laid in a shallow cup or pad of vegetation in a hole or crevice in a rocky area such as a cliff, ravine or cave, or in a wall of an abandoned building.

Range: Very large. Mediterranean eastwards to Japan and the Philippines. Native to nearly 90 countries.

Habitat: The Blue Rock-thrush is generally found on and around cliffs, gorges, ravines, rocky slopes and similar areas.

A beakful of tasty morsels – probably for a growing brood of youngsters. The Blue Rock-thrush eats mainly insects, such as grasshoppers, ants and flies.

Above: *As its name suggests, the Blue Rock-thrush is a species of mainly rocky, arid terrain, such as cliffs, ravines and crags. Despite males being a deep blue colour, it can be difficult to spot these birds even when they are pouring forth their melodious song.*

Right: *Female Blue Rock-thrushes are much duller than males. But both enjoy feeding on invertebrates such as spiders and earthworms and also small vertebrates such as mice and frogs.*

WHERE TO SEE The Blue Rock-thrush is a year-round resident and also breeds in Malta. It is a fairly common species and is best looked for in rocky coastal areas. A melodious songster, it is often initially seen sitting motionless on a rock or boulder, flying away when approached too closely.

The Blue Rock-thrush is one of the specialities found on the small island of Comino – a bird sanctuary and an IBA.

Cultural presence

Although protected by law, Blue Rock-thrush nestlings are often illegally taken and kept as caged songbirds. In fact, trapping wild birds for this purpose has long been a popular activity on Malta. A Life+ Project report on bird migration and trapping, published jointly by BirdLife Malta, the RSPB and Media Today in November 2009, points out: 'Most bird trappers love the birds they take from the wild and are not aware of any damage they may be causing to their natural environment. Since trappers do not deliberately kill birds, it is difficult for ordinary citizens to understand why trapping is not permitted under the EU's Birds Directive, while hunting (shooting) of certain species of wild birds outside the breeding period is.' According to the same report, Blue Rock-thrushes have been found dead in unattended trapping nets.

The Blue Rock-thrush has appeared on several Maltese stamps, including a set of bird stamps issued in 2001, and on coins.

The Blue Rock-thrush was depicted on two Maltese bird stamps – 5d and 10d – in 1971 as part of a four-value plant and birdlife set.

Source: *MaltaPost p.l.c*

MAURITIUS

Dodo

Raphus cucullatus

● EXTINCT

IUCN: It is not known precisely when the Dodo died out, although it is believed that the last birds disappeared sometime between 1662 and 1693.

Size: Length *c.* 70–75 cm (27.5–29.5 in).

Description: A dumpy, flightless bird that seemingly varied in colour from brownish or black to light grey and even white. The Dodo's featherless face was dominated by a large, yellowish bill. Males were bigger than females.

Diet: Nothing is known for sure about what the Dodo ate, although early mariners reported finding large gizzard stones on which they sharpened their tools. Gizzard stones indicate that hard seeds were a key part of the Dodo's diet, yet a Dutch sailor wrote in 1631 that the Dodo lived on raw fruits.

Reproduction: Unknown. Avian palaeontologist Dr Julian P. Hume, writing in his book *Extinct Birds*, cites a 1651 account by Cauche stating that the Dodo laid a single white egg in forests, but adds that this person did not actually land on Mauritius 'and appears to have confused his description with that of a Cassowary'.

Range: Occurred only on Mauritius, one of the Mascarene Islands. The Solitaire *Pezophaps solitaria*, a relative, lived on neighbouring Rodrigues.

Habitat: Recent research has revealed that the Dodo lived in forests dominated by Tambalacoque *Sideroxylon grandiflorum* trees and palms.

The real Dodo is sadly long gone, but the image of this, the world's most famous extinct bird, lives on in Mauritian life as the official national bird of Mauritius. It can be seen on the country's coat of arms (opposite a Sika Deer), on stamps and as a head-only three-dimensional watermark on banknotes. A commemorative set of 22-carat gold coins depicting the Dodo has been issued by the Central Bank of Mauritius.

There are also Dodo paintings, postcards, carved wooden toys and soft cuddly toys. Reunion Island has used the Dodo's name 'more ardently, even naming a beer after it'.

The large size of the Dodo's bill in relation to its head can be clearly seen in this photo of a skeleton.

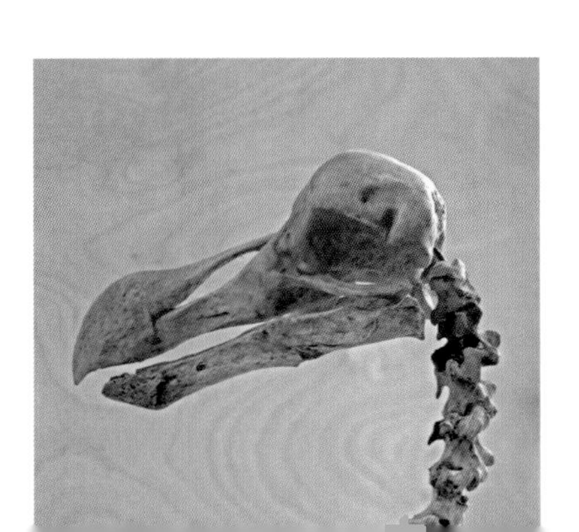

This coloured engraving from 1757 shows a Dodo and guinea pig for size comparison.

WHERE TO SEE A skeleton of a Dodo collected by barber Etienne Thirioux in 1900 is displayed in the Natural History Museum in Mauritius on the ground floor of the Mauritius Institute Building in Port Louis. The skeleton is unique in that all the bones are from a single bird.

A bronze replica of a Dodo – one of a number of replicas of extinct Mauritian animals – can be seen at the Ile aux Aigrettes nature reserve managed by the Mauritian Wildlife Foundation. Donated to the foundation, the aim of the replicas is to focus attention on 'the tragedy of extinction and the significance of our loss, and also to inspire a deeper understanding of the vital need for conservation'.

Cultural presence

No other country has adopted an extinct bird as its avian icon – but then no other country was once home to a bird whose name is now known the world over as a result of having been immortalised in the English language and literature.

The Dodo appears as a character in Lewis Carroll's much-loved children's book *Alice's Adventures in Wonderland*, while the commonly used expression 'as dead as a Dodo' has long been virtually synonymous with something that is out of date, dead or extinct.

The Dodo was one of 15 birds featured in a set of definitive stamps in Mauritius in 1965.

Although the Dodo's demise is often attributed to hunting by visiting mariners and early settlers, avian palaeontologist Dr Hume says in *Extinct Birds* that the introduction of Black Rats *Rattus rattus*, pigs, goats and possibly monkeys, 'all of which would have been direct threats to eggs and chicks and competitors for limited food resources, are the likely culprits'.

The Dodo became extinct during the Dutch period of colonisation of Mauritius. Dutch mariners dubbed it the Walghvogels, or nauseous bird, because its meat was tough to eat, although Admiral Jacob van Neck, who published the first written description of a Dodo in 1599, remarked that the cooked stomach and breast of the bird was 'extremely good'. It seems, however, that he and his fellow countrymen preferred eating Turtle Doves.

Dodo: The Bird Behind the Legend *by Alan Grihault tells the story of how humans discovered the Dodo, which had lived quite happily on Mauritius for thousands of years, and then wiped it out.*

MEXICO

Crested Caracara

NORTHERN CRESTED CARACARA

Caracara cheriway

The Crested Caracara is the unofficial national bird of Mexico, although a few sources claim it is the Golden Eagle.

WHERE TO SEE Common and widely distributed. The opportunistic Crested Caracara is never far away from carrion – either animals that have been knocked down by vehicles or have been killed by predators or died in some other way. It is equally at home ripping the flesh from large corpses, delicately picking maggots from rotting flesh and foraging on foot for invertebrates.

LEAST CONCERN

IUCN: Although precise numbers have not been ascertained, the population is very large and thought to be increasing.

Size: Length 51–64 cm (20–25.2 in). The biggest of all caracaras.

Description: Bare red face, large head and bill, black cap and crest and long yellow legs. Rest of body is brown-black/white with extensive barring. Four subspecies of Crested Caracara are usually recognised. *Caracara plancus audubonii* (often merged with *C. p. cheriway* according to volume 2 of the *Handbook of the Birds of the World*) is the subspecies found in Mexico.

Diet: Mainly an opportunistic scavenger, feeding on road kills and other carcasses. Harries vultures and forces them to regurgitate their food, catching it in mid-air. Also takes some live animals, from beetles, worms and nestling birds to freshwater turtles, crabs and snakes.

Reproduction: Generally two eggs are laid in a large, bulky and untidy nest in a tree, a cactus or even on the ground.

Range: Very large. Native to 27 countries, including Mexico.

Habitat: A bird of open and semi-open terrain. Often found on cattle ranches. Also frequents farmland, prairie, savannah, scrub and roadsides.

Biggest of all caracaras, the Crested has long, yellow legs and is able to walk and run with ease as well as soar and glide.

Above: *The Crested Caracara breeds from December to April in Central America, raising its young in an untidy nest of sticks, straw and such like which can be 100 cm or more across and up to 40 cm deep.*

Right: *A Crested Caracara carefully picking over the remains of a dead animal.*

Cultural presence

There is some confusion surrounding the identity of Mexico's national bird. Although the consensus is – at least as far as most online references are concerned – that the Crested Caracara is Mexico's national bird, some people maintain that the country's avian icon is actually the Golden Eagle *Aquila chrysaetos*. Unfortunately, even official Mexican sources contacted have been unable to give a definitive answer on this question.

Many national birds appear only in stylised forms when used on flags, currency, official seals and the like. Mexico's coat of arms is dominated by the image of what is clearly a large bird of prey. The raptor in question is showed perched on a Prickly Pear cactus, holding a snake in its bill and one of its talons. It is said that the Aztecs chose the site for their capital – the city of Tenochtitlan, founded in 1325 where Mexico City stands today – because that is where they observed an eagle grappling with a snake on a Prickly Pear cactus as seen by their leader in a dream.

The Crested Caracara is also known as the Mexican Eagle, despite the fact that it is not an eagle at all but a member of the falcon family. So does the bird on Mexico's coat of arms, which is apparently based on the Aztec legend, represent the Crested Caracara or, as some have suggested, a Golden Eagle? Certainly the bird depicted on the coat of arms – and on the Mexican flag as part of the coat of arms – looks more like a Golden Eagle than a Crested Caracara, but whether it is actually supposed to be this species is anyone's guess. It appeared in a set of stamps in 1996 dedicated to the protection of Mexican wildlife.

The Mexican coat of arms features a bird that looks more like a Golden Eagle, despite the national bird being a Crested Caracara.

Saker Falcon

Falco cherrug

At the time of writing, the Saker Falcon was the world's newest national bird, having been adopted as Mongolia's National Glorious Bird or National Bird of Glory by the government of that country in late 2012.

WHERE TO SEE The Saker Falcon can be seen potentially anywhere within its large range where there is suitable habitat, although there is a greater chance of seeing it in Mongolia or Tibet.

ENDANGERED

IUCN: Thought to be declining very rapidly as a result of unsustainable capture for falconry, habitat degradation and the effects of agricultural chemicals. The decline seems to be especially severe in central Asia. Globally it is estimated that there are 12,800 to 30,800 mature individuals.

Size: Length 45–55 cm (17.7–21.7 in). Females weigh considerably more than males.

Description: Stocky, solidly built falcon. Plumage is extremely variable – chocolate to sandy-brown with streaking.

Diet: Mainly small mammals, particularly rodents and lagomorphs (rabbits, hares and pikas). Also takes birds.

Reproduction: Usually three to five eggs are laid on a cliff ledge or crag, or in a tall tree.

Range: Very widely distributed – occurs from eastern Europe to western China. Breeds in many countries, including Mongolia.

Habitat: Essentially a bird of wild, open terrain (for example steppe, rocky areas and high plateaux). More varied habitat is occupied outside the breeding season.

Right: *The Saker is a superb aviator, taking prey completely by surprise at low level or by dropping like a stone and stooping on them from above.*

Far right: *A medium-sized bird of prey, the Saker is a bird of wide, open spaces where it hunts mostly small mammals, as well as birds, reptiles and insects. As illustrated here, the Saker usually looks for prey from a lofty vantage point.*

Cultural presence

Above: *The Saker Falcon was featured in a set of 12 falcon stamps issued in Mongolia in 1999. A set of six Saker stamps, proclaiming this species as national bird of Mongolia, was issued in June 2013.*

Below: *Falconry is an age-old sport and cultural practice in Mongolia. The Saker Falcon was even revered by warrior Genghis Khan.*

Now that this species has been formally adopted as Mongolia's national bird, one assumes the bird's image will be increasingly used in official and commercial circles.

The process of choosing an avian icon was started in 2009 by ornithologists from the National University of Mongolia and Mongolian Ornithological Society, including Dr Sundev Gombobaatar. A public survey revealed that just over half of all respondents felt that the Saker Falcon should be the national animal of Mongolia. Dr Gombobaatar has been quoted as saying that any national animal must have historical links with its country, pointing out that Mongols revered birds as animals of the sky, and that Chinggis Khaan and Khubilai Khaan both hunted with Saker Falcons. Dr Gombobaatar added that a national bird should symbolise a country's strength and glory, that it should not have been chosen by any other nation and that citizens should be educated about their avian icon.

Dr Gombobaatar is also quoted as saying that thousands of Saker Falcons are electrocuted by high-voltage power lines, and that a protective barrier should be erected around them. He has called for a nationwide survey of Saker Falcons to ascertain if this species is present in Mongolia in sustainable numbers.

According to a scientific paper entitled 'Developing a Sustainable Harvest of Saker Falcons (*Falco cherrug*) for Falconry in Mongolia', by Andrew Dixon, Nyambayar Batbayar, Gankhuyag Purev-Ochir and Nick Fox, Mongolia exported 3,141 Saker Falcons from 1997 through to 2010. The vast majority of these birds, 1,472, went to Kuwait. The annual number exported ranged from 25 to 402 birds. The authors say that the cost of a single Saker Falcon permit in 2010 was 11,760 US dollars, adding that 'the real benefit of the Saker Falcon trade to Mongolia is the connection it brings to leaders in Arabian states and the inward investment that can be derived from these relationships.'

Wrestlers warm up at Mongolia's national games by making flapping movements to imitate falcons like the Saker Falcon, and hawks.

The Saker Falcon has long been an integral part of the life, culture and history of Mongolia and was formally designated as the country's national bird on 20 October 2012 at the 63rd meeting of Mongolia's ministers.

It is reported that from now on the Saker will not only be promoted as a symbol of the country's culture, power and unity but also used to foster a love of the natural world. Furthermore, citizens have been called upon to help protect their special national bird from what has been described as 'unproductive loss' and contraband.

Montserrat Oriole

Icterus oberi

The Montserrat Oriole was declared to be Montserrat's national bird in 1982. It is 'a great source of local pride and fascination', according to the 2005–2009 Species Action Plan (SAP).

The Montserrat Oriole has appeared on badges, bumper stickers, posters and billboards, among other things, as a result of a major educational campaign aimed at developing a sense of pride in Montserrat's national bird and an understanding of and concern for this species and its habitat.

WHERE TO SEE Restricted almost entirely to Centre Hills – a lush, wildlife-rich area with steep-sided valleys. The Oriole Walkway trail through the Hills, a dormant volcanic cone, has been described as a birdwatchers' paradise. Experienced forest rangers can imitate the Montserrat Oriole's call, thus attracting these inquisitive birds to a very close range.

● **CRITICALLY ENDANGERED**

IUCN: The Montserrat Oriole has been driven to the brink of extinction by deforestation by humans exacerbated by repeated eruptions of the Soufrière Hills Volcano. Predation of nests by introduced rats *Rattus* spp. and the native Pearly-eyed Thrasher *Margarops fuscatus* is making a difficult situation worse. The number of birds left is estimated to be only 920 to 1,180 mature individuals.

Size: Length 20–22 cm (7.9–8.7 in).

Description: Males are striking birds with a black head, mantle, breast, tail, bill, legs and feet, but a yellow belly, lower back and rump. Females are yellow-olive.

Diet: Insects.

Reproduction: Lays two to three eggs in a hanging, basket-like nest.

Range: Found only on the island of Montserrat in the Lesser Antilles.

Habitat: Moist forests.

Right: *The female Montserrat Oriole is a drab yellow-olive.*

Far right: *Male Montserrat Orioles, which are predominantly black and yellow, are striking birds.*

Cultural presence

Although many birdwatchers visit Montserrat, the economic value of the oriole to the island has never been estimated, according to the 2005–2009 Species Action Plan (SAP). 'Many decision-makers are largely unaware of the important contribution the o riole and its habitat make to the Montserratian economy.' The SAP adds that if this species should become extinct, fewer nature-loving tourists might visit Montserrat, with adverse knock-on effects for the economy and employment in the tourism, guiding and conservation sectors.

Eruptions of Soufriere Hills Volcano from 1995 to 1997 destroyed two-thirds of the Montserrat Oriole's remaining habitat and virtually wiped out the species in the Soufrière and South Soufrière Hills. The bird is now confined to the Centre Hills and a small forest remnant in the South Soufrière Hills.

Two-thirds of the Montserrat Oriole's remaining habitat was destroyed by volcanic eruptions from 1995 to 1997.

The Montserrat Oriole's population in 2012 was thought to be 'stable but subject to strong annual fluctuations', according to Steffen Oppel, Senior Conservation Scientist with the RSPB. 'It seems that adults suffer high mortality in years with heavy ash-fall.' Some Montserrat Orioles have been taken into captivity as insurance against possible extinction in the wild.

An innovative educational campaign conducted in 1990 to raise awareness of and concern for the Montserrat Oriole used a variety of methods to get its conservation message across. They included puppet shows, youths dressed as orioles and heliconias (the national flower) parading through Montserrat's capital, Plymouth, jingles and songs such as 'Oriole of the Emerald Isle' by Ted Jordan, featuring Monty the Oriole who wears a 'suit of black and yellow', and a music video, slide shows, talks and even sermons. Among the more conventional promotional tools were posters ('to protect your national bird is to love our country'), badges, bumper stickers, billboards, a fact sheet and questionnaires.

The Montserrat Oriole has also been featured on stamps, is the logo of the Montserrat National Trust and is the name of a café near government offices in Brades.

The Montserrat Oriole was one of 13 species spotlighted in a set of bird stamps issued in 1970 on chalk-surfaced paper. The oriole stamp had a face value of 25c. In 1971, the Montserrat Oriole was again featured on a 25c stamp in a set of 10 bird stamps – this time issued on glazed paper.

African Fish-eagle

Haliaeetus vocifer

LEAST CONCERN

IUCN: Very large population of 300,000 mature individuals, which appears to be stable.

Size: Length 63–75 cm (24.8–29.5 in). Medium-sized fishing eagle.

Description: Adults are a striking combination of white, chestnut-brown and black. Long, broad wings, short, rounded tail and large head with prominent bill. The loud cry of an African Fish-eagle is one of the most distinctive bird sounds in Africa.

Diet: Mainly live fish, which it swoops down on and grabs from the water in its talons. Also takes nestling and even some adult birds. Piratical, harrying other birds such as storks, herons and kingfishers into dropping their prey.

Reproduction: Two eggs are usually laid in a large stick nest in a tree near water. Although sexes are similar in appearance, females are larger.

Range: Found over much of sub-Saharan Africa to the most southerly parts of South Africa. Locally common to uncommon or rare.

Habitat: Wide-ranging. Lakes, rivers, floodplains and stocked dams inland. Estuaries, creeks and mangrove lagoons in coastal areas. Fish-eagles spend most of their time perched in suitable habitats.

The African Fish-eagle has not always been Namibia's national bird. At one time the beautiful Crimson-breasted Shrike *Laniarius atrococcineus* occupied this position. It is said that the shrike was dropped on the grounds of national sensitivity because its colours too closely resembled those of the German flag (black, red and gold), Namibia having once been part of the former German Empire. Namibia became independent of South Africa on 21 March 1990.

The African Fish-eagle, it could be argued, is a more appropriate national bird for it is a far more conspicuous species and more widely appreciated by the people of Namibia. The African Fish-eagle is depicted, wings outspread, on Namibia's coat of arms, introduced when the country became independent, and said to be symbolic of the farsightedness of the country's leaders. The coat of arms, in turn, appears on the President's flag and on Namibia's national seal.

Given its iconic status, it is unsurprising that the African Fish-eagle's image is used in many spheres of life. For example, it is the emblem both of Namibia's national rugby union team (nicknamed the Welwitschias or Biltongboere) and of Fish Eagle Productions (a website, video, DVD and broadcasting business).

WHERE TO SEE The African Fish-eagle is one of Africa's most familiar birds and native to no fewer than 41 countries on the continent.

It is likely to be seen wherever there is water – everything from freshwater lakes, rivers and dams to swamps, creeks, estuaries, alkaline lakes and mangrove lagoons.

Top: *Mainly chestnut and white, the compact African Fish-eagle is a glorious sight to behold when in flight.*

Above: *African Fish-eagles spend most of their time perching on the lookout for prey.*

Cultural presence

The African Fish-eagle appeared on the back of a Namibian 5-dollar coin, and was one of various birds and other animals used to illustrate a set of stamps depicting the flora and fauna of the Caprivi area of Namibia. Designed by Mary Jane Volkmann, the Caprivi stamps won first prize in the geographical category of the 1999 Stamp World Cup competition in Paris as the most beautiful stamps in Africa and the Middle East, and third prize overall for the most beautiful stamps in the world. Volkmann's 60-cent stamp depicting the African Fish-eagle was also voted Namibia's most popular stamp in 1998.

Right: *The coat of arms of Namibia, and official heraldic symbol, features an African Fish-eagle flying above a pair of Oryx. It was introduced in 1990 at the time of independence.*

Below: *The Namibia Rugby Union is the governing body for rugby union in Namibia; its logo features an African Fish-eagle in flight.*

NEPAL

Himalayan Monal

IMPEYAN PHEASANT

Lophophorus impejanus

LEAST CONCERN

IUCN: Although the overall population has yet to be determined, it is thought that this species is common and widespread throughout its range wherever there is suitable habitat. However, it is declining as a result of habitat destruction and degradation, and overhunting for food.

Size: Length 63–72 cm (24.8–28.3 in).

Description: Adult males are striking iridescent birds with a metallic green head and crest, a black throat, copper, yellow-green and purple upperparts, velvet-black underparts and a chestnut tail. Females are dark brown with a white throat.

Diet: Tubers, seeds, shoots and berries, as well as insects and their larvae.

Reproduction: Three to five eggs are usually laid in a scrape on the ground.

Range: Very large. Native to Bhutan, mainland China, India, Myanmar, Nepal and Pakistan. This species is thought to have the greatest altitudinal range of any Himalayan pheasant.

Habitat: Meadows and clearings within coniferous and mixed forests where there is a well-established understorey of rhododendrons, bamboo and other plants.

Also known as the 'Impeyan Pheasant' (after Lady Mary Impey, wife of Sir Elijah Impey, a former Chief Justice of Bengal), the Himalayan Monal is not only the national bird of Nepal but also the state bird of Uttarakhand in India. It is featured prominently in local folklore and formed part of Nepal's old, pre-30 December 2006 coat of arms.

WHERE TO SEE Areas in which the Himalayan Monal can be seen include Khaptad National Park in western Nepal, Makalu Barun National Park in eastern Nepal and Rara National Park in the north-west of the

Adult male Himalayan Monals are a kaleidoscope of iridescent colours.

Females are largely dark brown birds

country. The Himalayan Monal is believed to undergo the greatest altitudinal movement of all Himalayan pheasants, according to the *Handbook of the Birds of the World* (volume 2). Research conducted in the late 1970s and early 1980s in north-west India revealed that monals lived at an altitude of 2,100 to 3,300 metres in December and January, but lower down, at an altitude of 1,700 to 2,900 metres, in February and March. In April, most monals moved from low altitude coniferous forests to high altitude oaks at an altitude of 2,000 to 3,500 metres. The birds continued to move upwards as spring progressed, the majority of monals being found above 2,700 metres during May – either in high altitude forests or in alpine meadows above the tree line.

Cultural presence

Leading ornithologist Dr Hem Sagar Baral of Bird Conservation Nepal wrote in a paper presented at the 4th International Galliformes Symposium in China in 2007 that all Nepalese tribes are familiar with the country's different pheasants, with each species having its own unique name. Such birds, he added, are often used in brand names or as logos. The *Danphe*, in fact, is the name of Bird Conservation Nepal's own bulletin.

Dr Baral says that the bird is wrongly named Himalayan Monal, for 'Monal or Munal actually means Tragopan to many Himalayan people. The name should be changed to Danphe, which is the Nepali word for the bird.'

He adds that this species is described in Nepali literature and folklore 'as the most colourful of birds – elegant looking and very majestic with attractive crown feathers. We say Danphe has nine major colours on its body. Its beauty, combined with its habitat in the High Himalayas, must be why the bird has been chosen as the national bird of Nepal.'

A song called 'Danphe and the Brain' appeared on Scottish band Mogwai's 2008 album *The Hawk Is Howling*. A flightless tropical bird called Kevin that appeared in the 2009 Disney/Pixar film *Up* was based on a male Himalayan Monal at Sacramento Zoo.

According to Bird Conservation Nepal, the Himalayan Monal is poached for its feathers and meat.

The Himalayan Monal appeared on two Nepalese stamps in 1959 and one in 1979. It was also depicted on a 25-rupee coin in 1974 and on the 50-rupee banknote in 2005 commemorating the Golden Jubilee of the Nepal Rastra Bank.

The Himalayn Monal was one of three birds featured in a set of Nepalese stamps in 1979 to mark the International World Pheasant Association Symposium.

NEW CALEDONIA

Kagu

Rhynochetos jubatus

An iconic bird, the Kagu is the national bird of New Caledonia and has a highly public profile within the country. It has appeared on many stamps during the past century or so, including various definitives (for example those issued in 1905, 1948, 1989 and 2005); four different-value WWF stamps issued in 1998; a BirdLife International stamp (part of a set of three) issued in 2007; and a Blue River National Park sheet issued in 2011.

The Kagu appears on the flag of the OPT Nouvelle-Caledonie (Post and Telecommunications Office government agency) and forms part of the logo of the Société Calédonienne d'Ornithologie (SCO), the BirdLife International affiliate in New Caledonia. Kagu souvenirs abound.

WHERE TO SEE Most Kagus live in Parc Provincial de al Rivière Bleue (Blue River National Park), where the population was estimated to be 500 birds in 2007.

ENDANGERED

IUCN: Only a 'very small, severely fragmented population' of probably 850-plus Kagus is left in the wild on New Caledonia, says BirdLife International. Although the species is declining overall, some groups of birds are growing in number as a result of conservation work.

Size: Length 55 cm (21.6 in).

Description: An unmistakable flightless, ash or pearl-grey bird endowed with a spectacular crest, which can be raised and opened in a fan-like display. Also striking are the Kagu's black-and-white-banded wings, its large, dark red eyes, and its orange-red bill and legs. It is the sole surviving member of the Rhynochetidae family.

Diet: Eats a variety of small animals including worms, millipedes, beetles, snails and lizards. Finds its prey mainly by foraging in leaf litter and soil. Kagu stand sentry-like until they detect prey by sight and/or sound. Invertebrates are sometimes exposed by brushing the leaf litter with their feet.

Reproduction: Each year one egg is laid on the ground in a simple nest of leaves, or in no nest at all. Most Kagu eggs are laid in July. Incubation takes 33 to 37 days. Both parents look after the chick.

Range: Found only on the island of Grand Terre in New Caledonia – a semi-autonomous French territory in the South Pacific.

Habitat: Both wet and dry forests, as well as certain shrub-covered areas. Kagus are agile birds and capable of moving long distances relatively quickly. An ascent of 750 metres in two days has been recorded.

The full beauty of the Kagu is revealed when it opens its wings to display its strongly patterned outer wing feathers.

Cultural presence

Kagu feathers were once used to adorn the war headdresses of the tribal chiefs of New Caledonia's indigenous Kanak people, and the Kagu's song was sung at Kanak war dances, according to the *Handbook of the Birds of the World* (volume 3). The Kanak also caught and ate Kagus, as did colonising Europeans.

After being described scientifically for the first time in 1860, Kagus were eagerly sought after by zoos and museums. Live birds were caught and sold as pets to Europeans.

The Kagu was almost driven to extinction in the 19th century as a result of being hunted for its highly prized crest feathers, which were incorporated into ladies' fancy hats.

Dogs have been a threat to the Kagu ever since they were introduced to New Caledonia by Captain James Cook in 1774. Other threats include forestry, mining, fires and introduced deer. The latter are severely damaging trees in the Boulouparis/La Foa, Canala triangle (one of the top Kagu areas outside Parc Provincial de la Rivière Bleue).

In 2008 the Société Calédonienne d'Ornithologie (SCO), with the support of partner organisations, launched the Kagu Recovery Plan (KRP), a key aim of which is to better understand the Kagu's distribution and population trends.

The traditional way of conducting a bird census is to listen for singing individuals at dawn. This method does not work with the Kagu, however, because these elusive, difficult-to-spot birds do not sing every day. As a result, several visits have to be made to the monitoring sites, many of which are hard to reach and therefore difficult to survey regularly. To overcome this problem, the SCO and Kagu Recovery Group have set up a network of song meters which are programmed to record for two hours daily at around sunrise. After seven days the meters are retrieved, the recordings are downloaded and the spectrograms are analysed for the presence or otherwise of singing Kagus. The meters are then moved to another site, and the process is repeated. Recordings have so far been obtained at more than 100 locations. Human listeners are also used at nine stations in Parc des Grandes Fougères. Eventually it is hoped to use song meters to record Kagus throughout New Caledonia.

This is one of three New Caledonian Kagu stamps issued as definitive in 1948.

NEW ZEALAND

Kiwi

Apteryx spp.

No other national bird is more closely associated with its country than the kiwi is with New Zealand – the official national bird. Even New Zealanders themselves are affectionately known as Kiwis the world over. The nickname was originally given to New Zealand soldiers during the First World War (1914–1918). Towards the end of this conflict, New Zealand troops at Sling Camp on Salisbury Plain in Wiltshire, England, carved a giant kiwi in the nearby chalk downs. Today all New Zealanders, irrespective of occupation, social background, age or gender, can be called Kiwis.

The kiwi is widely used in both official and unofficial circles in New Zealand. In some quarters the bird is represented as a straightforward silhouette, as on the roundel of the Royal New Zealand Air Force. In other walks of life it is depicted in a more stylised manner, as on the logo of the Bank of New Zealand Save the Kiwi Trust, or turned into a cartoon-like mascot like the one designed by Michael Tuffrey, a New Zealander of Samoan descent, for Radio New Zealand.

● VARIOUS

IUCN: There are five species of kiwi. The Great Spotted *Apteryx haastii* is classified as Vulnerable; Little Spotted *A. owenii* as Near Threatened; Northern Brown *A. mantelli* as Endangered; and Southern Brown (Tokoeka) *A. australis* as Vulnerable. The Rowi (Okarito Brown) *A. rowi*, which was designated as a separate species in 1994, is Under Review at the time of writing, but classified as Nationally Critical in New Zealand.

Size: The Great Spotted is the biggest kiwi at 45 cm (17.7 in) in length, and the Little Spotted is the smallest at 30 cm (11.8 in). The Northern Brown, Southern Brown and Rowi are all medium-sized kiwis measuring 40 cm (15.7 in) long.

Description: Although the different kiwis, which are largely nocturnal, differ from one another in size and plumage, all have the following unique set of characteristics in common. They are flightless with residual feathers variously described as resembling hair or fur; they have cat-like whiskers; their bones contain marrow, unlike those of any other birds; their nostrils are positioned at the end of the long, decurved bill, unlike those of any other birds; and their eggs are the biggest in the world in relation to their body size and contain more yolk.

Diet: Kiwis are omnivorous, although they eat mainly worms. Their diet can also include everything from woodlice, spiders, snails, slugs and grubs to seeds, berries and plant matter.

Reproduction: One large egg (approximately one quarter of the weight of the female) is laid in a burrow. Chicks hatch fully feathered and are soon independent of their parents.

Range: Kiwis are endemic to New Zealand. The Great Spotted is found only on the South Island (22,000 mature individuals, decreasing); the Little Spotted has been reintroduced to five offshore islands and one mainland site (1,200 mature individuals, stable); the Northern Brown is now found mainly on the North Island (35,000 mature individuals, decreasing); the Southern Brown is restricted to parts of the South Island (27,000 mature individuals, decreasing); and the Rowi is found mainly in the Okarito Kiwi Zone of the South Island (only about 375 birds left). The main threat facing most species is introduced predators such as the Stoat.

Habitat: Varies according to species (for example forests, grasslands, scrub and pasture).

Above: *Kiwis, which have an acute sense of smell, are the world's only known birds with nostrils at the end of their bills.*

Left: *A Kiwi lays one large egg in its burrow.*

Above: *Kiwis are nocturnal birds, emerging from their burrows at nightfall to forage for worms, insects, fallen fruit and such like.*

The most famous product to use a kiwi as its logo is the shoe polish of the same name. Although developed in Australia, creator William Ramsay named the polish after the kiwi in honour of his wife, Annie, who was born in New Zealand. Launched in 1906, Kiwi polish is now made all over the world.

WHERE TO SEE Kiwi Encounter at Rainbow Springs Kiwi Wildlife Park in Rotorua, on the North Island, is New Zealand's 'largest and most successful kiwi conservation centre', more than 1,000 captive-bred Northern Brown Kiwis having been released into the wild since 1995. Rainbow Springs' visitors can get up-close-and-personal with kiwis in the nocturnal house and outdoor night-time viewing enclosure.

Kiwis can also be seen in nocturnal enclosures at Otorohanga Kiwi House and Native Bird Park in Otorohanga, on the North Island, which boasts that it has New Zealand's 'largest public collection of native birds'.

Cultural presence

To the Maori the kiwi was the hidden bird of Tane Mahuta, God of the Forest. Clearly these indigenous New Zealanders knew that the kiwi was not just one bird but several closely related kinds because they had different names for them. For example, the Maori name for the Great Spotted Kiwi is Roroa or Roa, and for the Little Spotted Kiwi, Kiwi Pukupuku. The extremely rare Rowi is now usually known by its Maori name rather than by its former common name of Okarito Brown Kiwi.

Kiwis were a valuable resource for the Maori, but they were hunted sparingly, using dogs and traps, and with great ceremony. Birds were preserved in their own fat and cooked by steam in earth ovens, providing a valuable source of protein.

Kiwi feathers were used by the Maori to make cloaks – mainly for tribal chieftains – either by weaving the feathers into a flax fabric or by sewing together whole skins. Called Kahu-Kiwi, these garments 'carry the Wairua or spirit of the birds themselves', says the BNZ Save the Kiwi Trust website. 'At significant times – deaths, marriages and other great events – a Kahu-Kiwi is drawn over the shoulders as a privileged symbol of chieftanship and high birth. Today the tradition is continued using feathers gathered from kiwi that die naturally or through road accidents or predation.'

The BNZ Save the Kiwi Trust, a partnership between the Bank of New Zealand, Department of Conservation and Royal Forest and Bird Protection Society, seeks to protect kiwis and their habitats, and to ensure that the various species flourish in the future.

This distinctive road sign indicates the presence of kiwis in an area of New Zealand.

The Kiwi has appeared on both coins and stamps. In fact, New Zealand's currency is commonly know as the 'kiwi dollar'. The stamp shows the Southern Brown Kiwi and was one of eight, native bird-themed stamps issued in 1988.

The kiwi appeared on the 5-pound note issued by the Reserve Bank of New Zealand when it was established in 1934 and is currently featured on the Bank of New Zealand's 1-dollar coin. In fact, New Zealand's currency is often referred to as the 'kiwi dollar'.

This large corrugated iron sculpture of a kiwi is located in Otorohanga, New Zealand.

The kiwi has long been depicted on New Zealand stamps, from the 6-pence pictorial stamp of 1898 to the collectable Round Kiwi set of stamps reissued in 2011 and the 60-cent and 10-cent surcharge kiwi stamps also issued in 2011 to help raise funds for New Zealand's most vulnerable children.

Right: *The kiwi forms an integral part of the logo of the Bank of New Zealand.*

The famous Kiwi boot polish, launched in 1906, is known all around the world.

NICARAGUA

Turquoise-browed Motmot

Eumomota superciliosa

The Nicaraguan name for the Turquoise-crowned Motmot is *Guardabarranco*, which means 'ravine guard'. Singing duo Katia and Salvador Cardenal, formed in 1990, named themselves Guardabarranco after their country's national bird.

WHERE TO SEE This species can be seen in many parts of Nicaragua, especially in the Granada, Masaya, Carazo and Rivas departments, as well as in Managua, the capital city.

LEAST CONCERN

IUCN: The population could be as high as half a million mature individuals overall.

Size: Length 33–38 cm (13–15 in).

Description: A long, black-tipped, racquet-like tail is this bird's most striking feature. It is an exotic-looking species whose main plumage colours in both sexes are turquoise, green and brown. Other diagnostic features include a black facial mask and large bill.

Diet: Various insects such as butterflies, bees and dragonflies; spiders, worms and other invertebrates; small snakes and lizards; and some fruits.

Reproduction: Four eggs are usually laid in a long burrow, at the end of which is the nest chamber. The burrow is excavated in an earth bank or low cliff.

Range: Very large. Native to six Central American countries: Costa Rica, El Salvador, Guatemala, Honduras, Mexico and Nicaragua

Habitat: From woods and forests to thickets, plantations and gardens.

Cultural presence

The Turquoise-browed Motmot has appeared on various Nicaraguan stamps. The bird was also depicted on the reverse side of the country's 200-cordoba banknotes, issued in 2007.

A colourful Nicaraguan 'culture magnet', a promotional, fridge-type magnet, features the Turquoise-browed Motmot along with other Nicaraguan symbols, including the country's national flower, coffee bushes, lakes and volcanoes. The national plant is the colourful Sacuanjoche flower, produced by the *Plumeria alba* tree.

This Turquoise-browed Motmot stamp was one of seven bird stamps issued in 1989 to mark Brasiliana 89.

Black Crowned Crane

Balearica pavonina

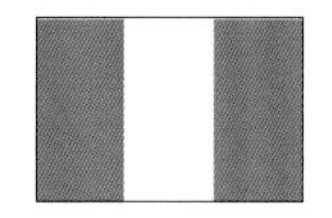

The Black Crowned Crane, with its distinct crown of stiff golden feathers in both male and females, is the official national bird of Nigeria. The image of this bird appears to be little used in Nigerian society.

WHERE TO SEE The Nigerian Crane Foundation says the places where the Black Crowned Crane are known to be present are not accessible for security reasons. They are easier to see in other countries.

VULNERABLE

IUCN: There are two subspecies, the one occurring in Nigeria being the West African Crowned Crane *B. p. pavonina*. Its total population has declined to an estimated 15,000 birds. In Nigeria, however, this subspecies is on the verge of extinction.

Size: Length 100–105 cm (39.4–41.3 in).

Description: Black head and a largely black body with some white, yellow and brown feathers. The single most eye-catching plumage feature is the crown of stiff golden feathers. The West African subspecies is distinguished from the East African by having red on only the lower half of its cheek patches, whereas the red on the East African subspecies extends into the upper half of the cheek patches.

Diet: Like other cranes, the Black Crowned Crane is omnivorous. Among its key food items are the tips of grasses, seeds from agricultural crops, insects and other invertebrates, and small vertebrates.

Reproduction: Several eggs are laid on a circular nest platform of grasses and sedges within or adjacent to thickly vegetated wetlands.

Range: The West African subspecies is found from Senegal to Chad in northern-central Africa.

Habitat: Frequents both wet and dry open areas, but prefers freshwater marshes, wet grasslands and the edges of water bodies.

The Black Crowned Crane is unlikely to be mistaken for anything else in flight.

Black Crowned Cranes prefer freshwater marshes, wetlands and the edges of lakes or pools, but can be found in dry habitats.

Cultural presence

Live trapping of birds for sale to local, regional and international markets 'for considerable profit' is the main threat facing the Black Crowned Crane in Nigeria, according to the International Crane Foundation (ICF).

There is 'considerable demand' for live cranes in North Africa, the Middle East and Europe, and exporting birds to these parts of the world is an important source of income for the various groups involved in this trade.

Despite the fact that the Black Crowned Crane was listed in Appendix 2 of CITES in 1992, which means that a licence is required to export the birds, 'there is strong evidence that international trade continues to deplete the wild population', according to the ICF.

Domestication of cranes in West Africa is encouraged by many local traditions, for these birds symbolise high social status, wealth, happiness, power and protection from evil spirits.

Black Crowned Cranes are opportunistically hunted for meat and for their body parts for use in traditional medicine, 'but cranes no longer occur in sufficient numbers to sustain the practice in many areas'.

A survey of the trade in Black Crowned Cranes in northern Nigeria in 2001 revealed that a seller in Kano market could expect to make a profit of 15,000 naira or 150 US dollars from each bird. Cranes are exported from Kano to Saudi Arabia, Qatar, Oman and other countries. Heads and feathers especially are used in traditional healing.

Black Crowned Cranes, says the ICF, are also threatened indirectly throughout their range by ineffective law enforcement, insufficient penalties for illegal activities, inadequate policies and legislation to protect key habitats, and a lack of educational programmes to stress the importance of saving wetlands.

Ruth Akagu of the Nigerian Conservation Foundation states that following the surveys supported by the International Crane Foundation and Wetland International in 2001, it has become clear that Black Crowned Cranes 'have been extirpated from where they once thrived. The fact that this species has been adopted as Nigeria's national bird has nothing, I believe, to do with its status but that the bird is revered in Nigeria.' She adds that it has not been possible to ascertain the Black Crowned Crane's current status because the areas in which it is known to be present are inaccessible for security reasons.

This crane stamp, issued in 1966, is said to show not a Black Crowned Crane but a Grey Crowned Crane.

White-throated Dipper

Cinclus cinclus

The White-throated Dipper was voted Norway's national bird in 1963.

WHERE TO SEE The White-throated Dipper is never far from fast-flowing rivers or streams with exposed rocks and weirs on which it perches between dives. This unusual bird can swim under water, using its wings as fins, and even walk on stream- and riverbeds.

LEAST CONCERN

IUCN: It is estimated that there are one to four million mature individuals globally and 170,000 to 330,000 breeding pairs in Europe.

Size: Length 17–20 cm (6.7–7.9 in).

Description: Nominate race is a dumpy, dark brown bird with a white bib. Sexes are similar. The subspecies of White-throated Dipper found in Scandinavia is the nominate race, *Cinclus cinclus cinclus*.

Diet: Aquatic insects and larvae, especially mayfly and stonefly nymphs and caddis-fly larvae.

Reproduction: Three to six eggs are laid in a large, domed nest with a side entrance and usually positioned near running water (for example by a mill, under a bridge, in a natural or man-made crevice or ledge, or even behind a waterfall).

Range: Huge. Native to no fewer than 63 countries, from Afghanistan to Austria, mainland China to Cyprus, Iraq to Ireland, and Ukraine to the UK.

Habitat: Mainly invertebrate-rich, fast-flowing streams and rivers with exposed rocks.

Cultural presence

The Norwegian national section of the International Council for Bird Preservation (now BirdLife International) and the Norwegian Broadcasting Company asked the public to suggest a suitable species in a series of five broadcasts from March to May of 1963. The dipper won with 1,182 of the 4,819 valid votes (24.5 per cent of the total) and was officially nominated in 1963.

Capable of diving, swimming under water and even walking on stream and riverbeds in search of prey, the charismatic White-throated Dipper is often associated with human settlements and structures in riparian environments, and generally enjoys a positive public image.

The White-throated Dipper was featured on one of four Norwegian bird stamps issued in 1980 with two booklets.

PAKISTAN

Chukar Partridge

Alectoris chukar

LEAST CONCERN

IUCN: The global population was estimated in 2004 to be around two million individuals.

Size: Length 32–39 cm (12.6–15.4 in).

Description: An attractive, boldly marked, dumpy bird, its distinguishing features being a black gorget, heavily barred flanks and a red bill and legs. Upperparts are brown/bluish-grey. Sexes are similar. There are 14 Chukar subspecies, the one found in Pakistan being *Alectoris chukar koroviakovi*.

Diet: Roots, grain, shoots and insects.

Reproduction: Generally seven to 12 eggs are laid in a scrape in the ground.

Range: The most widespread of all partridges, the Chukar is native to 31 countries – including Pakistan – and has been introduced to a number of others.

Habitat: Sparsely vegetated, arid, rocky terrain in Pakistan.

The Chukar is Pakistan's national bird. Its name is of Urdu origin – *Chakhoor.*

WHERE TO SEE The species can potentially be seen wherever there is suitable habitat. It is said to be widespread in less accessible areas.

Cultural presence

In the Punjab this species has been a symbol of love. Mythologically it is said to be captivated by the Moon and to be always gazing up at it.

In colonial India British sportsmen regarded the hunting of Chukar as good sport; fortunately this has not affected the Chukar population negatively.

The Chukar appeared on stamps in 2003 and 2009.

Like many young species of bird, this young Chukar appears scruffy in comparison to the adults.

This triangular Chukar stamp was issued in Pakistan in 2009 to mark National Environment Year.

Palau Fruit Dove

Ptilinopus pelewensis

LEAST CONCERN

IUCN: The population, according to BirdLife International, 'may be moderately small to large', and appears to be stable. There were an estimated 45,000 birds in 1991.

Size: Length 23–25 cm (9–9.8 in).

Description: Apart from a purple crown patch, the rest of the head, as well as the neck and breast, are grey. Orange belly band, olive-green back, wings and tail, grey bill and dark red legs.

Diet: Fruits. Feeds mainly in the canopy.

Reproduction: One egg is laid on a twig platform in a tree or shrub.

Range: A species with a restricted range, endemic to the island nation of Palau in the Pacific Ocean.

Habitat: Forests.

The Palau Fruit Dove was chosen as Palau's national bird by students in a nationwide election. The election itself was one of the first tasks of the Palau Conservation Society, which was formed in 1994 when the island nation became an independent republic.

WHERE TO SEE The Palau Fruit Dove can be found in wooded areas of the Palau archipelago, which consists of around 340 islands and islets. According to *Pigeons and Doves: A Guide to the Pigeons and Doves of the World*, the Palau Fruit Dove is known from Babelthuap, Koror, Peleiu and Angaur islands.

The Palau Fruit Dove is as exotic and as colourful as the environment in which it lives.

The purple crown patch and olive-green back stand out in this picture of the Palau Fruit Dove.

The Palau Fruit Dove appeared on this 20c stamp in 1983.

Cultural presence

The Palau Fruit Dove's image has graced many Palau stamps over the years, and it was one of 11 bird species featured in a set issued in 1991. It is also the logo of the Palau Conservation Society.

Also known as the Biib, the Palau Fruit-dove is described by the Palau Conservation Society on its website as 'one of Palau's most loved endemic birds'.

According to legend, the Palau Fruit-dove was once a beautiful young woman named Biib, 'who radiated grace and beauty', and who loved a handsome young man called Matkerumes. Biib was forced to marry Chief Osilek of Ulong, but she 'resented and hated him unceasingly'. When Biib contracted leprosy, Osilek stayed away from her. Even Biib's own mother could not stand the sight of her daughter's badly disfigured face and body.

After Matkerumes bathed Biib in a waterhole containing *ditmechei* (a plant endowed with magical power by the gods), she miraculously recovered and her beauty was restored. As the sweethearts set off together in a canoe, Biib's mother drowned herself, calling to her daughter: 'Look at me for the last time, because once I dive down I will turn into a clam. My dear daughter, remember never to eat clam.' Seeing this, Biib also committed suicide in the same manner.

Says the Palau Conservation Society: 'If one has a Biib as a pet bird, he should not eat clams or bring them into the house, because they were once Biib's mother and the bird will die.' It adds that the sad cooing of the shy bird can be heard in Palau's forests.

The Palau Conservation Society logo features a Palau Fruit Dove.

PALESTINE

Palestine Sunbird

Cinnyris osea

● **LEAST CONCERN**

IUCN: The global population of the Palestine Sunbird has yet to be ascertained, but appears to be stable. Although it is uncommon throughout much of its range, it is locally common to abundant.

Size: Length 8–10 cm (3.1–3.9 in).

Description: Males appear to be dark-coloured birds when seen from a distance. Close-up and in the right light, the iridescent, multicoloured (largely blue-green, purple and black) glory of these sunbirds becomes apparent. Decurved black bill and black legs. Females are mainly a nondescript grey-brown. There are two subspecies of Palestine Sunbird *Cinnyris osea*, the one found in Palestine being the nominate race of *C. o. osea*.

Diet: Nectar from many flowers, as well as fruits, seeds, spiders and insects.

Reproduction: One to three eggs are laid in a pear-shaped nest built solely by the female, usually in a bush.

Range: Very large. Native to 15 countries or territories, including Palestine.

Habitat: From dry grassland, riverside bushes and rocky areas to parks, gardens and orchards.

The Palestine Sunbird was not the official bird of the Palestinian Authority at the time of writing, but could well become so in the near future. Moves are afoot to seek approval for the adoption of this popular bird as Palestine's official avian emblem.

Palestinian artist and photographer Khaled Jarrar has designed an unofficial 'State of Palestine' passport stamp and also a 750-fils (the old Palestinian currency) unofficial postage stamp. The former features a stylised image of a Palestine Sunbird, whereas the latter, which Jarrar had printed privately, incorporates a photograph of the species.

WHERE TO SEE Palestine Sunbirds are just as likely to be seen in flower-filled gardens as in more natural surroundings. 'You do not really need to go far from a city to meet a sunbird,' says the Palestine Visitor Information Centre. 'House gardens full of flowers are like perfect restaurants where they gather to sing and eat.'

The Palestine Sunbird usually feeds while perched alongside flowers, but it can also hover as it takes nectar and insects.

Cultural presence

Left and centre: *Centrepiece of the Sunbird Pavilion at the British Council's International Architecture and Design Showcase 2012 was a towering figure of a Palestine Sunbird made from two laser-cut timber structures.*

The Palestine Sunbird is 'a national and folk symbol' for Palestinians, according to the Palestine Visitor Information Centre in Bethlehem. It is a widely painted species, and one delightful work of art featuring it is by the late Ismail Shammout (1930–2006), a Palestinian refugee and artist.

The British Council's International Architecture and Design Showcase 2012, held at the Dreamspace Gallery in London, incorporated a Palestine Sunbird Pavilion, the aim of which was to showcase recent projects in the West Bank and Gaza Strip by the Palestine Regeneration Team. The centrepiece of the Pavilion was a huge Palestine Sunbird made from laser-cut timber – a familiar species in a region noted for being one of the world's most important bird-migration routes. The Palestine Sunbird, it would seem, was also used to symbolise freedom of movement and hope. The British Council said at the time that the Palestine Sunbird's 'physical structure and busy animation suggests a freedom of movement and a hopefulness that seem to be far beyond the control of any enforced borderlines'.

The Eagle of Saladin depicted on Palestine's coat of arms symbolises independence. Saladin (1138-1193) was a Muslim of Kurdish origin – and the first Sultan of Egypt and Syria – who led Islamic opposition to the European Crusaders. The discovery of an eagle on the wall of a citadel built by Saladin in Cairo has led some people to believe that this bird was the Sultan's personal symbol.

The Palestinian coat of arms incorporates the Eagle of Saladin.

PANAMA

Harpy Eagle

Harpia harpyja

An open-winged eagle endowed with a crest just like that of the Harpy Eagle has long formed part of the Republic of Panama's coat of arms. But precisely what eagle species was this heraldic bird meant to represent? All was eventually made clear when a law was passed in 2006 declaring that the heraldic eagle was indeed the Harpy Eagle. It was actually made Panama's national bird by another law, in 2002.

WHERE TO SEE The Harpy Eagle's largest known population is in Darien National Park, where the Peregrine Fund is working to conserve the species and its habitat through environmental education, increased involvement by local communities and basic research aimed at understanding this species' population ecology.

A life-size nest, information panels and a video can be seen at a Sony-sponsored Harpy Eagle exhibit at Summit Botanical Garden and Zoo within Soberania National Park.

NEAR THREATENED

IUCN: The number of mature individuals is estimated to be less than 50,000 overall. It is thought that this species is undergoing a moderately rapid decline as a result of hunting and habitat loss.

Size: Length 89–105 cm (35–41.3 in); wingspan up to 2 m (6.6 ft); weight 7.6–9 kg (16.8–19.8 lb) females, 4–4.8 kg (8.8–10.6 lb) males. A huge bird of prey – one of the biggest and most powerful in the world.

Description: A strikingly plumaged raptor. Sexes are similar. Adults are largely a combination of black, white and grey. Grey head and neck (head appears bigger when the nape crest is raised); dark breast band; dark upperparts; white underparts; long, barred tail; massive bill; huge talons – hind claw is up to 7 cm (2.8 in) long.

Diet: Monkeys and sloths especially, but also other tree-dwelling and terrestrial mammals, reptiles and sizeable birds, including curassows and macaws. Fast, agile hunter, seizing many prey animals as it flies through the forest. It is also an ambush predator by rivers and salt licks.

Reproduction: Two eggs are laid in a very large stick nest high in a tree, but only one chick is raised. Harpy Eagles breed infrequently – every third year according to the *Handbook of the Birds of the World* (volume 2).

Range: Occurs in Central and South America, from Mexico to Argentina, but is thinly distributed and generally rare.

Habitat: Lowland tropical forests.

Cultural presence

The Harpy Eagle is aptly named, for 'Harpy' is derived from the Greek word for snatcher, and this species is certainly adept at snatching monkeys, sloths and other animals while flying stealthily through the trees. In Greek mythology Harpies were half-bird, half-woman winged deities that snatched food, objects and even people.

Since becoming Panama's national bird, considerable efforts have been made in Panama to increase public awareness of the Harpy Eagle and an understanding of the threats it faces in the wild. Panamanians celebrated FestiHarpia or Harpy Eagle Day at Summit Botanical Garden and Zoo on 10 April 2011.

The Audubon Society of Panama and Harpy Eagle Friends Foundation produced 5,000 bookmarks of one of Canadian wildlife artist David Kitler's Harpy Eagle paintings to raise the profile of this species among students.

David spent a month with his wife, Ly, in Panama's Darien Province in December 2005. During this time he conducted the most comprehensive artistic study to date of the Harpy Eagle, in support of conservation efforts to restore the species to its historical ranges and to boost public awareness about the key role this apex predator plays in a healthy ecosystem.

David's trip was made possible by a grant from Artists for Conservation. A second trip to the Darien Province has since been made to deliver donated school supplies and baseball equipment to the village of Llano Bonito, where David and Ly were based for much of the time during their 2005 trip. They were the first outsiders to stay in Llano Bonito, where Embera tribal people live as their ancestors did through subsistence farming, hunting and fishing. Baseball is very popular among the villagers, and David and Ly hope the baseball equipment will provide a recreational option for children in Llano Bonito and other nearby villages who have often distracted themselves by persecuting local wildlife, including the Harpy Eagle.

Harpy Eagles are killed for food and for their feathers, or out of fear. Tail feathers are used in ceremonial headdresses, and shamans use feathers in their rituals. An adult female Harpy Eagle was shot in Panama by villagers, according to Artists for Conservation, 'because they were afraid that such a large and powerful bird might attack and eat their children, even though no such attacks have ever been recorded'.

Artist David Kitler and his wife Ly with tribal people in Panama. The couple have done much to raise awareness of the Harpy Eagle in this country in support of conservation efforts.

One of David Kitler's works of art depicting the Harpy Eagle.

Raggiana Bird-of-paradise

Paradisaea raggiana

Also known in New Guinea as the *Kumul*, the Raggiana Bird-of-paradise (named after the Marquis Francis Raggi of Genoa) became Papua New Guinea's national bird in 1971.

Raggiana Bird-of-paradise's image appears on the country's flag as a yellow silhouette with its wings open and plumes trailing on a red background above five stars of the Southern Cross constellation. On the national crest it is shown displaying its wings and plumes while perched on a *kudu* (drum) and spear.

LEAST CONCERN

IUCN: Although no one knows how many Raggiana Birds-of-paradise there are in the wild, it is thought that this species is common and that the population is stable.

Size: Length 34 cm (13.4 in).

Description: Like the dozens of other members of the exotic bird-of-paradise family, the Raggiana Bird-of-paradise is breathtakingly beautiful. Males are endowed with a yellow head and collar, dark emerald throat, bluish-grey bill, blackish upper breast, salmon-pink to fiery orange flank plumes, and a pair of long black tail 'wires'. Female birds, by comparison, are rather drab.

Diet: Fruits (it is the main disperser of seeds for some mahogany and nutmeg species) and arthropods (for example insects, spiders and crustaceans).

Reproduction: Usually two eggs are laid in a bowl-shaped nest on a branch 2–11 m (6.6–36 ft) above the ground. Two to 10 males at a time compete for the attention of females in traditional, communal lekking trees. The birds a dopt a variety of showy poses and postures to display their red flank plumes to best effect.

Range: Distributed widely in southern and north-eastern parts of Papua New Guinea (PNG).

Habitat: Mainly moist lowland forests. Other suitable areas include degraded former forest areas and rural gardens.

It is depicted on stamps, coins and banknotes, as well as on the aircraft of Air Nuigini (Papua New Guinea's national airline). It has also appeared on flags of the former Australian (British) Trust Territory of Papua and New Guinea.

According to *Birds of Paradise: Revealing the World's Most Extraordinary Birds*, by Tim Laman and Edwin Scholes, the Raggiana Bird-of-paradise's likeness is widely used on both domestic and export products.

Above: *A vision of beauty in paradise – a male Raggiana, like Joseph, wearing his coat of many colours.*

Below: *A lekking Raggiana Bird-of-paradise bending over to impress an approaching female with his red flank plumes.*

WHERE TO SEE One of the best places to see birds-of-paradise in Papua New Guinea is the Baiyer River Wildlife Sanctuary, north of Mount Hagen, the capital city of Western Highlands Province.

Cultural presence

Birds-of-paradise have long occupied a special place in the tribal cultures of Papua New Guinea, the highly prized feathers from these exotic birds being incorporated into headwear and other items of clothing, especially those of a ceremonial nature. Explorer and scholar Antonio Pigafetta of Venice, who is said to have been one of the first Europeans to set eyes on birds-of-paradise in the early 1600s, recorded in his diary that the native people he encountered told him that such birds originated in a paradise on Earth and were known as *bolon diuata* (birds from God).

Because the first birds-of-paradise to arrive in Europe had been mounted without legs or wings, it was initially believed that they were magical creatures that flew eternally among the clouds (which is why, according to the legend, they did not need legs), lived on cloud dew, obtained their colours from flying close to the sun and fell to Earth when they died.

During the 19th century there was such a huge demand for bird-of-paradise feathers by designers making fashionable ladies' hats that bird numbers began plummeting. By the early 20th century, according to one source, no fewer than 50,000 birds-of-paradise were killed and traded annually. Following an export ban in 1922 they started to make a comeback.

Although commercial exploitation of birds-of-paradise is banned, some are still sold locally – mainly for their feathers – according to Miriam Superna, Co-director of the Papua New Guinea Institute of Biological Research. Surveys by the Institute have revealed that while some Papuans buy or rent traditional adornments for their celebrations, others use ones that have been handed down from generation to generation and kept in bamboo tubes or suitcases.

Feather plumes from Raggiana and other birds-of-paradise are used by the hunter-gatherer Huli wigmen of Papua New Guinea's Tari Highlands to adorn their elaborate and highly colourful headwear. They are also noted for a bird dance in which they mimic the movements of birds-of-paradise found in the area in which they live.

Tribal people in Papua New Guinea have long incorporated bird-of-paradise feathers in their headware.

Papua new Guinea's Air Nuigini has adopted the Raggiana Bird-of-paradise as its logo.

Papua New Guinea's national rugby league team is nicknamed the Kumuls after this bird-of-paradise species and the team's logo features a yellow bird-of-paradise. Rugby league was first played in Papua New Guinea in the late 1940s; it was introduced there by Australian soldiers.

The Raggiana Bird-of-paradise appeared on this 1984 Papua New Guinea definitive.

Bare-throated Bellbird

Procnias nudicollis

The Bare-throated Bellbird was chosen to be Paraguay's national bird as a result of a public vote. Known locally in Paraguay as the *Guyra Pong,* this species is more often heard than seen.

WHERE TO SEE The tropical humid forest of UNESCO's Bosque Mbaracayu Biosphere Reserve – the first reserve to be designated as such in Paraguay.

VULNERABLE

IUCN: This species is believed to be declining rapidly as a result of deforestation and widespread trapping for the cage-bird trade. It is thought that there are no more than 10,000 mature individuals left in the wild.

Size: Length 25.5–31 cm (10–12.2 in).

Description: Males are striking birds, for they are completely white, except for their greenish-blue eye patches and black bill. Females are drab by comparison.

Diet: Fruits only.

Reproduction: Breeding data is sparse. One of the few nests ever found was a small, cup-like structure made from epiphyte rootlets. It is thought that a single egg is laid per clutch.

Range: Occurs only in Brazil, Paraguay and Argentina.

Habitat: Atlantic rainforest.

Cultural presence

The Bare-throated Bellbird appeared on a 2004 Paraguayan stamp and is the emblem of FAUNA Paraguay – an online community whose aims include promoting the conservation and protection of Paraguay's fauna and habitats.

In fact, it is an aptly named bird, for it produces loud, metallic sounds. One of its calls can be heard up to 3 km (2 miles) away. In some ways it is surprising that this species is prized as a cage bird, because the noises it makes are distinctly unmusical.

According to Paul Smith of FAUNA Paraguay, the Bare-throated Bellbird is 'quite culturally ingrained in Paraguayan folklore – as are many birds which often have an indigenous Guarani legend attached to them'. One famous Guarania song, 'Pajaro Campana', incorporates the species' song into its music, which is typically played on a harp.

The Bare-throated Bellbird is listed by FAUNA Paraguay as one of the 10 most sought-after species as far as visiting birdwatchers are concerned.

The online community and conservation group FAUNA Paraguay has a Bare-throated Bellbird on its logo.

Andean Cock-of-the-rock

Rupicola peruvianus

The Andean Cock-of-the-rock is the national bird of Peru. It is also one of the country's most sought-after and widely photographed birds. It has appeared on various Peruvian stamps. It is has a striking plumage and unique mating rituals. In the mating season, males gather to 'lek' or perform a dance to impress female onlookers. The best dancer usually chooses first.

The female Andean Cock-of-the-rock has a smaller crest than the male and is largely orange-brown plumage.

LEAST CONCERN

IUCN: Although the global population has yet to be ascertained, it is thought to be stable. This species is said to be uncommon and patchily distributed. There are four subspecies, three of which are found in different parts of Peru, including the nominate race, *R. p. peruvanius*, which occurs in central Peru.

Size: 30–32 cm 11.9–12.6 in).

Description: Males are bizarre-looking yet stunningly beautiful birds and the closest thing South America has to birds-of-paradise. Except for their wings and tail, which are black, males have a vivid orange or scarlet plumage. The most eye-catching feature is the semi-circular crest. Females are a distinctly less striking orange-brown in colour and have a smaller crest. The Andean Cock-of-the-rock should not be confused with the Guianan Cock-of-the-rock *Rupicola rupicola* – a completely separate species.

Diet: Mainly fruits, but also large insects. Nestlings enjoy a more omnivorous diet, of which frogs and small lizards appear to be important constituents.

Reproduction: Males are renowned for their spectacular mating or lekking displays, during which they excitedly bob, bounce and bow, flap their wings, snap their beaks, and utter a variety of grunting and squawking sounds. Two eggs are laid in a cone-shaped mud nest, generally on a rock face.

Range: Occurs in five South American countries – Bolivia, Colombia, Ecuador, Peru and Venezuela.

Habitat: Montane forests.

Left: *A female Andean Cock-of-the-cock and young in a nest.*

Below: *Andean Cock-of-the-rocks – considered to be the Neotropical equivalents of birds-of-paradise – make a variety of non-musical noises, including a pig-like foraging/flight sound.*

WHERE TO SEE One of the best places to see this species is the aptly named Cock-of-the-rock Lodge, which is situated at an altitude of 1,600 m (5,250 ft) in a cloud forest 165 km (103 miles) from Cusco on the eastern slopes of the Andes. A noted location for watching and hearing lekking male Andean Cocks-of-the-rock, the lodge supports and protects a 5,060-hectare (12,500-acre) cloud forest reserve.

This Andean Cock-of-the-rock stamp was issued in 1972.

The male Andean Cock-of-the-Rock is a bizarre-looking bird thanks to its unmistakable crest, the feathers of which form a semi-circular casque.

Cultural presence

The Andean Cock-of-the-rock is probably better known outside than inside Peru, given the international popularity of birdwatching. Peru is renowned for the richness of its avifauna and the Andean Cock-of-the-rock is one of the most sought-after species among globetrotting birders and therefore a valuable tourism resource. The Peruvian Cock-of-the-rock has appeared on Peruvian stamps on at least three occasions – in 1972 (one of five species featured on a set of Peruvian bird stamps); in 2003 (two different Cock-of-the-rock stamps); and in 2004 (one stamp to mark National Biodiversity Day).

PHILIPPINES

Philippine Eagle

MONKEY-EATING EAGLE

Pithecophaga jefferyi

The Philippine Eagle was declared to be the Philippines' national bird by presidential proclamation on 4 July 1995, even though it was already the country's official avian icon. The new proclamation, by President Fidel V. Ramos, was made because before that date the common name for this species was not the Philippine Eagle, but the Monkey-eating Eagle. The latter was itself proclaimed to be the national bird of the Philippines in 1978 by President Ferdinand E. Marcos.

● **CRITICALLY ENDANGERED**

IUCN: It is estimated that there are only 180 to 500 mature individuals left in the wild. The population has declined rapidly during the past half-century. Deforestation, slash-and-burn agriculture and uncontrolled hunting are among the threats facing this species.

Size: Length 86–102 cm (33.9–40.2 in). One of the world's biggest, most powerful and rarest eagles.

Description: A majestic raptor with a huge bill, dark face, shaggy, mane-like crest, dark brown upperparts and creamy-white underparts.

Diet: Mainly mammals, especially Flying Lemurs *Cynocephalus volans* and Common Palm Civets *Paradoxurus hermaphroditus*. Also preys on other birds, such as Rufous Hornbills *Buceros hydrocorax*, as well as reptiles.

Reproduction: The Philippine Eagle pairs for life. A single egg is laid in a large platform nest of sticks and twigs 27–50 m (88.5–165 ft) from the ground on a substantial branch or in the fork of a tree. Incubation lasts from 58 to 68 days. Young birds fledge when they are four to five months old but stay in the vicinity of the nest for nearly a year and a half. Captive birds have lived for more than 40 years.

Range: Endemic to the Philippines. Found only on Luzon, Samar, Leyte and Mindanao islands – mainly on the latter.

Habitat: Dwells mainly in primary dipterocarp (large, long-lived, mostly evergreen tropical trees) forests but also frequents secondary and gallery forests. In Mindanao each Philippine Eagle pair on average covers an area of 133 sq km.

Cultural presence

A stylised white-strokes image of the Philippine Eagle forms the centrepiece of the new logo of Bangko Sentral ng Pilipinas (BSP) – the central bank of the Republic of the Philippines. BSP chose the bird because it is a symbol of 'strength, clear vision and freedom', the qualities to which the bank aspires.

The Philippine Eagle has appeared on a variety of stamps over the years, including a WWF set of four issued in 1991 (it was also featured in other WWF sets in 1997 and 2004) and on 50-centavo coins. Not surprisingly, it is also the emblem of the Philippine Eagle Foundation.

The non-profit Philippine Eagle Foundation – the nation's leading organisation for the conservation of birds of prey, which survives on donations – is dedicated to saving the Philippine Eagle and its rainforest habitat. Based near Davao City, the Foundation believes that the fate of the Philippine Eagle, the health of the natural environment and the quality of life for Filipinos are inextricably linked. It is therefore committed to work for the survival of the eagle as a wild species, 'the biodiversity it represents and the sustainable use of our forest resources for future generations to enjoy'.

The Foundation's wide-ranging programmes include the captive breeding of Philippine Eagles (the first two birds, Hope and Unity, hatched in 1992), preparing birds for release, rehabilitating injured wild birds, field research, community-based initiatives and conservation education.

The Foundation's Philippine Eagle Centre – a Filipino educational resource as well as a major tourist attraction – is home to a mixture of captive-bred and wild Philippine Eagles, as well as other bird and non-bird species. The Foundation's conservation staff organise tours of the centre, field trips, school visits and public exhibitions.

Philippine Eagle Week, which takes place from 4 to 10 June every year, spotlights the importance of the Philippines' national bird as a flagship species and barometer of biodiversity.

The Philippine National Bank (PNB) features a bird that resembles a Philippine Eagle.

Left: *The Philippine Eagle Foundation undertakes a wide range of work, including preparing birds for release and rehabilitating injured wild birds.*

Below: *Philippine Eagles are bred in captivity by the non-profit Philippine Eagle Foundation.*

PUERTO RICO

Puerto Rican Spindalis

STRIPE-HEADED TANAGER

Spindalis portoricensis

LEAST CONCERN

IUCN: This tanager species is described as common and widespread, although the number of mature individuals has yet to be established.

Size: Length 16.5 cm (6.5 in).

Description: Males are dapper, multicoloured birds. The diagnostic features are a black head and a white stripe above and below each eye. Neck and breast are yellow-orange, back is green, and wings and tail are black. Females are much duller than males – olive-green with streaked underparts.

Diet: Mainly fruits and flower buds.

Reproduction: Two to four eggs are laid in a cup-like nest low down in a bush or tree.

Range: Endemic to Puerto Rico, where it is resident. Also a vagrant to the Virgin Islands (UK).

Habitat: Forests, plantations and shrubby areas.

The Puerto Rican Spindalis is the national bird of this Caribbean island. Found throughout Puerto Rico, it is readily seen and a 'must' for visiting birdwatchers, given that it is native to no other part of the world.

WHERE TO SEE Anywhere there is suitable habitat, including gardens.

Cultural presence

Puerto Rican fashion designer Marta Negron's love of this beautiful bird led her to name her line of girls' occasional dresses after the Reina Mora (the species' local name).

The Puerto Rican Spindalis is sexually dimorphic, meaning that males and females have completely different plumages. Males are brightly coloured and stand out well in foliage, whereas females are rather drab by comparison.

A fine male Puerto Rican Spindalis. Its diagnostic feature is its black head, across which run two prominent white stripes.

SAMOA

Tooth-billed Pigeon

Didunculus strigirostris

Officially adopted as the national bird of Samoa, the Tooth-billed Pigeon is currently the subject of a government recovery plan seeking to ensure that this species is no longer threatened with extinction, to secure populations on the main Samoan islands of Upolu and Savi'i and to return the bird to many different forest areas.

The Tooth-billed Pigeon, which is known locally as the *Manumea*, was also adopted as the official mascot for the 13th South Pacific Games, held in Samoa in 2007.

WHERE TO SEE There are six IBAs in Samoa where the Tooth-billed Pigeon is present: O Le Pupu-Pu'e National Park on the island of Upolo; Central Savai'i Rainforest on the island of Savi'i; Apia Catchments, Upolo; Aleipata Marine Protected Area, Upolo; Eastern Upolo Craters, Upolo; and Uafato-Tiavea Forest, Upolo.

ENDANGERED

IUCN: Threatened by deforestation for agriculture, cyclones (devastating storms have destroyed or damaged many native trees), encroachment by highly aggressive, non-native trees and illegal hunting. There has been a dramatic decline in this species' population. Precise numbers are unknown, but there could be as few as several hundred birds left in the wild. An extensive population survey was due to take place in 2012.

Size: Length 31–38 cm (12.2–15 in).

Description: A dark, secretive bird. Largely glossy blackish-green and chestnut-brown, with red legs. The heavy, hooked and notched bill, from which the bird's name derives, is red (at the base) and orange-yellow. Sexes are similar but females are duller than males.

Diet: Frugivorous. Especially likes fruits of *Dysoxylum* genus trees in the mahogany family, using its specialised bill to open the tough, woody outer capsule, or pericarp.

Reproduction: Very little is known about this species' breeding behaviour.

Range: Endemic to Samoa.

Habitat: Occurs in and around the edges of mature native forests. Tends to stay within forests, rather than flying above them.

The Tooth-billed Pigeon is a strong, agile flier. Its wings make a loud clapping sound when flushed from vegetation.

Cultural presence

The Tooth-billed Pigeon was a popular choice for national bird. A marketing project found that 84 per cent of questionnaire respondents initially supported making the Tooth-billed Pigeon Samoa's national bird. However, by the end of the one-year marketing exercise, support had risen to 96 per cent.

There are various suggested origins for the name *Manumea* – the local name for Tooth-billed Pigeon. *Manu* means bird in Samoan, while *mea* may indicate that the bird is beautiful, decorative or bold, or simply be a reference to its colour.

At least some tribal elders attribute the origin of the name to the fact the Tooth-billed Pigeon is bigger than other pigeons, has a distinctive bill and a fiercer cry and strength.

The Tooth-billed Pigeon Recovery Plan, published by the Ministry of Natural Resources and Environment (MNRE) and spanning the period 2006 to 2016, states that it wants 'most Samoans to recognise the Manumea as a key part of their natural heritage and to play their part in its conservation'. The recovery plan has a variety of objectives, including managing key forest areas on Upolo and Savai'i (five sites on the former and three on the latter), eliminating shooting, 'which still occurs, even though it is fully protected', and establishing new populations on rat-free islands, at mainland sites and in captivity. Many aspects of the Tooth-billed Pigeon's ecology are unknown, so there are plans to conduct research to learn more about this species. Another objective calls for public awareness and education programmes.

The Tooth-billed Pigeon was one of 10 species featured in a set of Samoan bird stamps issued in 1967.

The recovery plan says the 20 or so priority actions 'will go a long way towards giving Manumea a long-term future'.

It states that the Tooth-billed Pigeon is very important to Samoa and its citizens. Traditionally, it was a highly regarded source of food, especially for high chiefs. According to one source (Appleton 1871), if a travelling party from a dominant tribe arrived at a village of a lower tribe where they intended to spend the day, 'they would order the chief man of the village to procure them a certain number of *Didunculus* before night. If they failed to provide the birds, a severe cudgelling would be the consequence.' According to another source (Stair 1897), the Tooth-billed Pigeon was once hunted for food in great numbers and caught either with bird lime or shot with arrows.

Today, it is claimed, the Tooth-billed Pigeon is shot purely accidentally by hunters while they are looking for other, better-tasting pigeons. The recovery plan, however, says this is questionable.

The recovery plan goes on to say that Tooth-billed Pigeon is the only bird endemic to Samoa widely known beyond the island nation, 'largely because early biologists described it as being closely related to the extinct, flightless Dodo, based on similarities between their beaks. Indeed, the Latin name *Didunculus* means "little dodo". It is now generally recognised that this similarity is an adaptation to feeding on large fruits that evolved separately in the two species. However, DNA analyses have shown the Dodo to be in fact a form of pigeon that most likely evolved from forms that flew to Mauritius from Africa.' The Tooth-billed Pigeon, says the recovery plan, is the one species that visiting birdwatchers most want to see.

The species plays a vital ecological role in Samoan rainforests by distributing seeds of native trees.

SCOTLAND

Golden Eagle

Aquila chrysaetos

LEAST CONCERN

IUCN: Stable population.

Size: Second biggest bird of prey in the UK after the White-tailed Eagle *Haliaeetus albicilla*. Length 75-90 cm and wingspan 190 to 227 cm, according to the *Handbook of the Birds of the World* (volume 2). Females are roughly 10 per cent bigger and up to 50 per cent heavier than males. Sexes are virtually inseparable in the field.

Description: Plumage predominantly dark brown to black-brown although the crown and nape are paler than the rest of the body – straw to tawny coloured. Massive bill, long wings and powerful, 4 to 6 cm long talons.

Diet: Various medium-sized mammals, including rabbits, hares, rodents, squirrels, small deer, young sheep and goats and sometimes also birds (prefers grouse, pheasants and partridges), reptiles, insects and other creatures. Frequently feeds on carrion.

Reproduction: Lays one to three eggs in a large, bulky eyrie of sticks and branches on a cliff ledge, crag or in a tree.

Range: Extremely large. Native to dozens of countries, including the UK. The global population is estimated to be around 170,000 individuals, but the RSPB says there are only 442 breeding pairs in the UK.

Habitat: In Scotland, the Golden Eagle frequents open moorland, mountains, remote glens and certain islands.

The Golden Eagle became the world's newest national bird in 2013 – albeit an unofficial one – when it was chosen by members of the public as Scotland's favourite animal. Following the Big 5 campaign, in which this raptor secured nearly 40 per cent of the popular vote, it is likely that the Golden Eagle's image will be increasingly used in official and unofficial circles. There are various subspecies, the one occurring in Scotland being the nominate race, *A. c. chrysaetos*.

WHERE TO SEE Eagle-watching areas in Scotland include the Beinn Eighe, Abernethy, Rum and Invereshie and Inshriach National Nature Reserves, and the isles of Mull, Skye and Islay.

Cultural presence

The Golden Eagle was voted as 'unofficial' national bird in 2013. Thousands of votes were recorded online during Scotland's Big 5 campaign – run jointly by Scottish Natural Heritage and VisitScotland as part of the Year of Natural Scotland 2013 celebrations and funded by both bodies as well as the Scottish Government.

The number of votes cast from spring to autumn was 12,417. Of these, 4,773 votes were for the Golden Eagle (nicknamed the High Flyer), 2,523 for the Red Squirrel, 1,819 for the Red Deer, 1,794 for the Otter and 725 for the Harbour Seal. The top three animals among the remaining 783 votes were Scottish Wild Cat, Pine Martin and Puffin. Voting closed on 31 October 2013.

'While we can be enormously proud of all our native wildlife, it is fitting that the magnificent Golden Eagle has topped this poll of Scotland's Big 5 species,' said Paul Wheelhouse, Scotland's Environment and Climage Change Minister.

He went on to say that all of the UK's breeding pairs of Golden Eagles are resident in Scotland and that there is still a need to protect them from illegal persecution occurring in some areas. In the last two centuries the species was nearly wiped out, but the population has recovered. He said that Scotland has a responsibility to protect these birds for future generations.

Scotland's Big 5 were selected because they are all high-profile species with a broad geographical spread and widely associated with the country. The Big 5 were championed through a series of billboard advertisements in Manchester, Liverpool, Newcastle, Glasgow and Edinburgh. TV personality Neil Oliver lent his voice to an extensive radio campaign with written press and online content specially developed throughout 2013.

Scottish Natural Heritage Chief Executive Ian Jardine said the response to the Big 5 campaign had been 'brilliant'. He added that several alternative lists had been put forward for seabirds, game animals and other plants, and there had also been support for rarer species such as the Pine Marten and Scottish Wild Cat.

'It has got people thinking and talking about wildlife and showing how much affection and pride people have not just for the five species on the list but for Scottish wildlife generally.'

VisitScotland Chairman Mike Cantlay remarked that the Big 5 campaign had been 'a cornerstone of our celebrations for the Year of Natural Scotland 2013. One of the key pillars of the year was to get as many people here at home out seeing parts of Scotland they may not have been to before.' The Big 5 campaign will hopefully encourage people to explore the remoter areas of Scotland.

In the Scottish Big 5 campaign vote, the Golden Eagle received the highest number of public votes, followed by the Red Squirrel, Red Deer, Otter and Harbour Seal.

Seychelles Black Parrot

Coracopsis nigra

Also known as *Kato Nwar* in Creole, the *barklyi* subspecies of the Black Parrot is one of the world's rarest birds and the national bird of the Seychelles. It is something of an icon on Praslin Island, and the Black Parrot Hotel and Black Parrot Suites of the Coco de Mer Hotel have been named after this bird.

WHERE TO SEE The Seychelles Black Parrot can be seen in the Vallée de Mai Nature Reserve in Praslin National Park on Praslin Island. Vallée de Mai was designated a World Heritage Site by UNESCO in 1983 as an outstanding example of low and intermediate altitude palm forest characteristic of the Seychelles. The forest is preserved in something approaching its primeval state.

The Seychelles Island Foundation (SIF) says on its website that Vallée de Mai is 'the exclusive home of the rare Black Parrot. There are only approximately 100 to 200 pairs of these birds in existence which thrive on the fruits and flowers of the mature palm forest. recent research has focused on the conservation and ecology of these fascinating birds and on vegetation studies.'

The species' stronghold is a 19.5-hectare (48-acre) ancient palm forest which boasts the world's biggest population of Coco-de-mer *Locoicea maldivica* trees. The Coco-de-mer is famous for being the bearer of the world's biggest nut.

LEAST CONCERN

IUCN: Although the species as a whole is officially classified as being of Least Concern, *C. n. barklyi*, or the Seychelles Black Parrot – one of four subspecies – is unofficially regarded as being Critically Endangered as a result of habitat loss, extreme isolation and introduced rats, which eat its eggs and chicks.

Size: Length 35–40 cm (13.8–15.7 in).

Description: Essentially a dark brown bird with a pale bill. The outer parts of the main flight feathers of *C. n. barklyi* are bluish-grey and the crown is lightly streaked. Sexes are alike. What the Seychelles Black Parrot lacks in terms of colour it makes up for in sound, producing flute-like whistles.

Diet: Palm and other fruits, including berries, and flowers and seeds.

Reproduction: Two or three eggs are laid in a tree cavity or in one of the rat-guarded nestboxes provided.

Range: The Seychelles Black Parrot is native to the Comoros, Madagascar and the Seychelles, but the *barklyi* subspecies is found only on Praslin and Curieuse Islands in the Seychelles. It is thought that fewer than 100 pairs are left on Praslin. *Barklyi* breeds on Praslin, but only feeds on Curieuse.

Habitat: Forests, scrub and gardens.

Cultural presence

The World Parrot Trust states that no one knows exactly how many Seychelles Black Parrots are left in the wild. 'It may be in serious trouble. Because it is found only on Praslin and Curieuse Islands, its loss would be a strong jolt to the area's ecosystem, which depends on the parrots for seed dispersal of local plants.'

The Trust has launched an appeal for funds to support a study by UK researcher Ellen Walford who, along with the Seychelles Islands Foundation, is conducting 'critical work determining accurate population accounts and learning about the availability of food and nest sites'.

These studies will take place on Praslin and Curieuse Islands.

The Seychelles Black Parrot or Lesser Vasa Parrot has appeared on various Seychelles stamps over the years, including a four-value set issued in 1976 to mark the 4th Pan African Ornithological Congress, and another four-value set issued in 1989 with island birds as its theme. Its image has also been used on the obverse of a 25-cent coin..

SIF

seychelles islands foundation

Logo of the Seychelles Islands Foundation, which manages and protects Vallée de Mai and Aldabra.

This Black Parrot stamp was one of four bird stamps issued in an 'island birds' set in the Seychelles in 1989.

Crimson Sunbird

Aethopyga siparaja

The Crimson Sunbird was chosen as Singapore's unofficial national bird by the public following an informal poll conducted by the non-profit, non-governmental Nature Society (Singapore) during its Nature Day celebrations in 2002.

The Crimson Sunbird is a small bird; the eye-catching male having a prominent crimson-coloured head, neck, throat and back. The female has an olive-green back and yellowish breast.

LEAST CONCERN

IUCN: Although the overall size of the population has yet to be established, this species is usually said to be common.

Size: Length 11.7–15 cm (4.6–5.9 in) males,10 cm (3.9 in) females.

Description: The upper breast, mantle and most of the crown on males of the nominate race are scarlet, the forehead to the centre of the crown is glossy purple-green and the rest of the underparts are olive-grey. Females are largely olive. Both sexes have a strongly decurved bill. This species and Vigors' Sunbird (*A. vigorsii*) are often considered to be the same species. There are various subspecies. The Crimson Sunbird subspecies found in Singapore is the nominate race, *Aethopyga siparaja siparaja*.

Diet: Insects, spiders and nectar.

Reproduction: One to three eggs are laid in a finely woven, pear-shaped nest.

Range: Extremely large. Native to 15 countries, including mainland China, India, Indonesia, Malaysia, Philippines and Singapore.

Habitat: Wide-ranging, from forests and scrub to plantations, coconut groves, parks and gardens.

Far left: *Adult male Crimson Sunbird in all its feathered finery. Non-breeding males lose most of the crimson on their heads and breast.*

Left: *Female Crimson Sunbirds are anything but crimson – just a plain olive-green.*

They have a thin down-curved bill and a brush-tipped tubular tongue specially adapted for nectar feeding. The long tongue extends well beyond the length of the bill. They supplement their diet with insects and spiders.

Crimson Sunbirds can hover in flight, but only for a second or two. It is possible they can feed and hover if they need to, but usually they perch to feed. Their flight is fast and direct.

WHERE TO SEE The Crimson Sunbird is found in forest and cultivated areas throughout Singapore. Specific localities where the Crimson Sunbird can be seen include the Bukit Batok Nature Park, Bukit Timah Nature Reserve, Central Catchment Nature Reserve, Pasir Ris Park, Pulau Tekong, Pulau Ubin, Sentosa, Singapore Botanic Gardens and Sungei Buloh Wetland Reserve.

They are known to be partial to feeding on Snakeweed, Torch Ginger, heliconias, Noni and Sea Apple.

Crimson Sunbirds love the nectar produced by plants.

Cultural presence

In the voting process for national bird, six bird species were nominated by the society for the public's consideration: White-bellied Sea Eagle *Haliaeetus leucogaster* (majestic); Racket-tailed Drongo *Dicrurus paradiseus* (elegant); Black-naped Oriole *Oriolus chinensis* (bright); Brown-throated Sunbird *Anthreptes malacensis* (home); Crimson Sunbird *Aethopyga siparaja* (tiny); and Asian Fairy-bluebird *Irena puella* (jewel). The Crimson Sunbird received the highest number of votes.

Organised by Nature Society Council member Sunny Yeo, the poll was held to stimulate interest in nature and wildlife. Alan OwYong of the Nature Society said the former Indonesian head of state, President Habibie, had coined the phrase 'Little Red Dot' for Singapore. 'The Crimson Sunbird is tiny but bright and energetic and best represents the island state.' Alan added that the Crimson Sunbird is only the unofficial national bird because it is not endorsed by the Singapore Tourism Board.

The Crimson Sunbird appeared on 5-cent Singapore stamps in 2007/2008 and in a 2013 Our City in the Garden set. It was also used on a decal by the Nature Society (Singapore) during one of its bird races.

In 2002 the Crimson Sunbird won a Nature Society (Singapore)competition for the unofficial title of Singapore's national bird.

SOUTH AFRICA

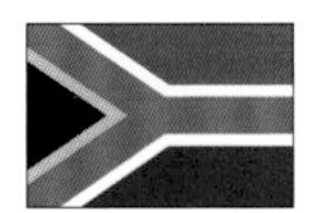

Blue Crane

STANLEY CRANE

Anthropoides paradisea (Grus paradisea)

VULNERABLE

IUCN: Overall, it is estimated that there are 26,000 mature individuals in the wild. Numbers have declined rapidly as a result of power-line collisions, poisoning, persecution, illegal trade and habitat loss. Nationally, the population has halved since the 1970s.

Size: Length 100–120 cm (39.4–47.2 in).

Description: Dainty, long-legged, bluish-grey crane with a bulbous, white-capped head and long, decurved 'tail'.

Diet: Omnivorous diet of seeds, roots, tubers, various invertebrates (for example locusts, grasshoppers and termites), fish, frogs, reptiles and small mammals.

Reproduction: Usually two eggs are laid in vegetation or mammal dung, or on the bare ground.

Range: Found almost only in South Africa.

Habitat: Natural grassland, pastures and agricultural fields (both crop and fallow).

The Blue Crane is the national bird of South Africa. Photographically and stylistically, its image has been – and still is – used widely in South Africa. It has appeared on various stamps, including a joint issue in 2004 featuring national birds of eight countries in southern Africa and a set of 'big five' South African birds in 2008; on 5-cent coins; on limited edition 2-rand coins commemorating the 50th anniversary of the WWF in 2011; and on educational posters. The 2002 Sasol SciFest, which focused on the plight of the Blue Crane, featured the bird on its poster and T-shirt.

The Blue Crane also adorns the logos of, for example, the Kwande Private Game Reserve, the Overberg Crane Group and Indwe Risk Services. It forms the crest of the South African Institution of Civil Engineering's coat of arms. Strangely, though, it is the Secretarybird *Sagittarius serpentarius*, not the Blue Crane, that appears on South Africa's national coat of arms.

WHERE TO SEE The Blue Crane is a partial migrant, undertaking seasonal movements within South Africa. It occurs in many parts of the

Found in many parts of South Africa, the graceful Blue Crane undertakes seasonal movements within the country. It is the most range-restricted of all the world's cranes.

Chicks, which are fed initially on insect larvae and worms, are able to fly when they are three to four months old. Parent cranes are very protective of their young.

country, especially the Western Cape, although numbers vary widely from area to area.

There are four tourist Blue Crane Routes in the Western Cape: Caledon or Overberg (where the Indian and Atlantic Oceans meet), Xairu (centred on the town of Heidelberg), Porterville (named after this town) and Southernmost. Kwande Private Game Reserve, near Grahamstown in the Eastern Cape, is also noted for its Blue Cranes. In fact, Kwande is the Xhosa name for Place of the Blue Crane.

Cultural presence

Blue Cranes figure prominently in the history, culture and traditions of South Africa. *Indwe* is the Xhosa name for this bird and *Indwa* the Zulu name. The Blue Crane is one of two species – the other is the *Igwalagwala* or Knysna Turaco *Tauraco corythaix* – long revered by the Zulu monarchy. It is said that before King Shaka (*c.* 1787–1828) came to power, Zulu kings wore a Knysna Turaco head feather. King Shaka, however, preferred a Blue Crane head feather, as does the current Zulu monarch, His Majesty Goodwill Zwelithini. A controversial multimillion-rand statue of King Shaka in warrior pose, showing him wearing a towering Blue Crane feather, was commissioned in 2011 from artist Peter Hall for Durban's new, state-of-the-art King Shaka International Airport.

Xhosa men who distinguished themselves in battle, as well as in other ways, were often presented with Blue Crane feathers by their chief at a special ceremony called *ukundzabela*. Such men wore the prized feathers in their hair.

Indwe is a village in the Eastern Cape and also the name of South African Express Airways' monthly inflight magazine.

The Overberg Blue Crane Group, a joint initiative between the Overberg community and CapeNature, is committed to conserving the Blue Crane in the Overberg and Swartland regions of the Western Cape where some 12,000 birds – half of the country's total population – make their home.

Blue Crane feathers have long been prized by Zulus in South Africa.

The Endangered Wildlife Trust, like the government, is keen to see DNA testing used to establish the parentage of all captive cranes in South Africa, according to Kerryn Morrison, Manager of the African Crane Conservation Programme. The African Association of Zoos and Aquaria has identified a studbook keeper to develop and maintain a sustainable captive population of Blue Cranes.

OVERBERG CRANE GROUP
OVERBERG KRAANVOËLGROEP

The Blue Crane is the logo of the Overberg Crane Group.

SRI LANKA

Sri Lanka Junglefowl

Gallus lafayetti

Formerly called Ceylon Junglefowl during the colonial era, today this species is known locally as *Walikukula*, and is the official national bird of Sri Lanka.

WHERE TO SEE The best places to look for this shy bird are Sri Lanka's national parks and forests, such as the Sinharaja Forest Reserve (a national park, biosphere reserve and Natural World Heritage Site), and the best times to look for it are at dawn and dusk.

Cultural presence

The Sri Lanka Junglefowl does not appear to have a high profile. It was, however, featured on Ceylon stamps in 1966 and 1967. The naming of the Sri Lanka Junglefowl commemorates the French aristocrat Gilbert du Motier, marquis de Lafayette.

The Sri Lankan Junglefowl appeared on one of four Birds of Ceylon stamps issued in 1966.

LEAST CONCERN

IUCN: Although the size of the Sri Lanka Junglefowl's population has yet to be determined, this species has been described as locally plentiful.

Size: Length 66–72 cm (30–28.3 in) males, about 35 cm (13.8 in) females.

Description: Colourful and elegant male Sri Lanka Jungefowl have a rusty-red body, glossy black wings, a yellow-centred red comb and red legs, and could be confused with domestic roosters. The much smaller females appear largely brown with spots and bars and are far less conspicuous.

Diet: Seeds, berries, flower petals and invertebrates. Often seen foraging on forest tracks early and late in the day.

Reproduction: Usually two eggs are laid on or just above the ground.

Range: Endemic to Sri Lanka. Found in many parts of the island.

Habitat: Forests, scrub, bamboo thickets and plantations.

Looking very much like a domestic rooster, the cock Sri Lanka Junglefowl sports a yellow-centred red comb and bare red face.

St Helena Plover

WIREBIRD

Charadius sanctaehelenae

Known locally as the 'Wirebird', the St Helena Plover is the national bird of St Helena. It adorns the coat of arms and also appears on the flag of this UK Overseas Territory.

It has appeared on various St Helena stamps over the years, including a set of five special St Helena Plover stamps in 2002. It was featured on the 5-pence coin issued in 1984 and subsequently appeared on other coins as part of the coat of arms.

St Helena Football Association, St Helena Cricket Association and St Helen Golf Club all use the St Helena Plover on their badges. One football team is even called the Wirebirds. It also appears on the flag of Harford Primary School at Longwood. The wharf arch bears a large plaster cast of the St Helena Plover.

The St Helena Plover was once often featured on inlay work on boxes, tables and other wooden objects, but this craft is little practised today.

WHERE TO SEE The St Helena Plover may be seen in lower altitude areas of St Helena, wherever suitable habitat exists over extensive flatter areas. Thirty-one breeding areas are censused each year, but birds also occur in other small pockets of habitat. Three of the key breeding areas are the Deadwood Plain pasturelands in the north-east, Man and Horse in the south-west and the arid Prosperous Bay area.

CRITICALLY ENDANGERED

IUCN: On average, 350 birds have been counted in the annual censuses during the past few years. Threats to the species include habitat changes and predation by cats, rats, the Common Myna *Acridotheres tristis* and sheep.

Size: Length 15 cm (5.9 in).

Description: Dark, medium-length bill; dark, longish legs; distinctive eye and face stripes, buff below with darker, buff-fringed feathers above. Sexes are similar.

Diet: Little is known about this species' diet, although insects (especially beetles) and snails have been recorded.

Reproduction: One to two eggs are usually laid in a scrape on the ground.

Range: Found only on St Helena. It is the island's sole surviving terrestrial endemic bird.

Habitat: Mainly flat, short-sward pastures and open arid areas with short herb vegetation.

A St Helena Plover chick starting to make its way in the world.

Cultural presence

The St Helena Plover enjoys a high profile on the island. The Friends of St Helena have even named their magazine after this bird.

Motion-sensitive infrared cameras have revealed high levels of nest disturbance and predation. Funding attracted by the long-running St Helena National Trust Wirebird Conservation Programme, supported by the RSPB/Defra and OTEP/DfID, has enabled a predator-recording system to be established. Thirty-two camera traps are being used to monitor the presence of cats, rabbits, rats and other animals. The traps are moved to different locations every fortnight. Data is also obtained on a monthly basis using tracking tunnels that record activity by rodents and other small, ground-dwelling animals. Species using these peanut-butter-baited tunnels leave telltale footprints on an ink-covered card.

A public-communication exercise, involving house-to-house visits, leaflets, newspaper articles and radio broadcasts, is keeping islanders informed about the various measures being taken and how they and their pets will be affected.

St Helena National Trust states that although few people 'will object to reducing the rat population, many people are rightly concerned about the intention to reduce the number of cats'. The Trust points out that the Wirebird Conservation Programme has no intention of harming domestic cats, even though they venture into the plovers' areas where they consider the birds to be either food or toys. The aim is to ring-fence important breeding areas and reduce predator numbers.

The flag of St Helena features a St Helena Plover.

A Species Action Plan launched in 2008 (a revised version was due to be published in 2013) seeks to boost the species' population over a 10-year period so that it can be downgraded to Vulnerable.

Dr Chris Hillman, Wirebird Conservation Programme Manager with St Helena National Trust, said in 2012: 'We are currently engaged in a number of activities focused on improving Wirebird numbers. The main problem identified has been the loss of eggs in nests (up to 80 per cent to both feral and domestic cats, as well as to rats and myna birds. There is also some loss at the chick stage to the same predators. We are concentrating our research therefore into the density and localities of cats and rats especially at four key locations. Ideally, this would be done island-wide to eradicate feral cats, but budgets are insufficient and St Helena is a very complex island terrestrially with thick bush and deep, ravine-like valleys. At the same time, we are monitoring Wirebirds and their nesting and fledging success to see what effect, if any, our control efforts are having.'

Dr Hillman added that building an airport on St Helena 'is leading to a raft of other developments, all of which require land and improved infrastructure, etc, and at some point are likely to have a knock-on effect on the Wirebird, not least its habitat. So, our work is cut out in trying to mitigate these effects.'

St Helena's coat of arms incorporates an image of the St Helena Plover.

The St Helena Plover was depicted on this 1975 12p stamp.

Brown Pelican

Pelecanus occidentalis

The Brown Pelican is the national bird of Saint Kitts and Nevis. Two stylised Brown Pelicans are depicted on the Saint Kitts and Nevis coat of arms – one on either side of the shield.

WHERE TO SEE Ponds and bodies of water of the south east Peninsula on Saint Kitts.

LEAST CONCERN

IUCN: It was estimated in 2009 that there are 300,000 mature individuals globally. Overall the population seems to be increasing.

Size: Length 105–152 cm (41.4–59.8 in), plus a cavernous, 280–348 cm bill.

Description: The only one of the world's seven pelican species to have a dark brown body. White head and neck and black legs and feet. There are six subspecies. The Brown Pelican subspecies found in Saint Kitts and Nevis is the nominate race, *Pelecanus occidentalis occidentalis*.

Diet: Mainly fish, especially anchovies and sardines. A plunge diver and the only pelican to regularly feed in this fashion.

Reproduction: Usually three eggs are laid in a shallow depression on the ground, but also nests in bushes and small trees.

Range: Huge. Native to more than 40 countries/territories from Canada to Chile, Ecuador to El Salvador and Peru to Puerto Rico.

Habitat: Shallow coastal waters such as estuaries and bays. The only essentially marine pelican.

Cultural presence

A common bird in Saint Kitts and Nevis, the Brown Pelican has adapted well to living alongside people and is often seen around fishing boats and fishing ports, where it quickly snaps up discarded fish or parts of them. The national bird was also featured on two stamps in 1972.

Above: *Although they appear ungainly birds, Brown Pelicans can glide with ease just above the waves.*

Left: *The Saint Kitts and Nevis coat of arms incorporates two Brown Pelicans.*

St Lucia Amazon

ST LUCIA PARROT

Amazona versicolor

Known locally as the *Jacquot*, the St Lucia Amazon was declared St Lucia's national bird in 1979. St Lucians started to care about their parrot and protection and education programmes have helped to save the species.

WHERE TO SEE The Government Forest Reserve in the central mountain massif is a major stronghold for the St Lucia Amazon. It also forages from time to time in the upper reaches of the Mandele Dry Forest.

VULNERABLE

IUCN: It is estimated that there are now around 500 wild birds. Numbers are slowly increasing as a result of protection laws and conservation work.

Size: Length 43 cm (17 in).

Description: Largely green, but with a blue face and forehead, a red/maroon breast patch, dark blue primary feathers, orange eyes and a grey bill.

Diet: Feeds on a wide range of fruits, seeds and foliage.

Reproduction: One or two eggs are laid in a cavity in a tall, mature tree.

Range: Found only on St Lucia. It is in fact the only parrot on the island.

Habitat: Prefers moist forests at an altitude of 500–900 m (1,640–2,952 ft).

The St Lucia Amazon is a bird of mountain forests.

Cultural presence

The St Lucia Amazon declined almost to the point of extinction as a result of habitat loss caused by deforestation, and uncontrolled hunting for food and the pet trade. By the late 1970s this species' entire global population had been reduced to a mere 100 birds or so all living in an area of just 60 sq km (23 sq miles).

A captive-breeding programme was started by the Durrell Wildlife Conservation Trust (DWCT) at Jersey Zoo in 1976 with nine wild-caught birds (seven fledglings and two adults) brought in from other collections. The programme chalked up its first breeding success in 1982. Seven years later, in 1989, St Lucia's Prime Minister took a pair of Jersey-bred birds back to St Lucia to be used in a second captive-breeding programme. 'For St Lucian people, these parrots – housed in aviaries funded by the Trust – are an opportunity to meet their national bird face to face and may be the only ones that some will ever see,' says the DWCT. 'Wild birds are very shy and retiring and their natural habitat is largely inaccessible to people. But they can be seen on bus tours that take ecotourists to the forest.' Field research and monitoring on St Lucia was supported by DWCT from 1975 to 1996.

The St Lucia Amazon appeared on one of four bird stamps issued by the island in 1969.

An imaginative public-education programme called Protection Through Pride utilised the Jacquot Express – a brightly painted bus provided by the World Parrot Trust that doubled as a classroom – to tour St Lucia's schools and villages. Jacquot himself – a person dressed up as a parrot – put in an appearance while children were taught about the St Lucia Amazon and its forest home. The original idea came from Paul Butler of RARE. 'This bus travelled all over the island, visiting schools and other locations, telling the story of the endangered St Lucia Amazon and what had to be done to save it from extinction,' said the World Parrot Trust.

"Our team at Paradise Park in the UK, led by David Woolcock and Nick Reynolds, bought a used bus and fitted it out with working models, video programmes and other educational facilities, then shipped it out on a banana boat and handed it over to the staff of the island's forestry department. It was a great success and resulted in similar buses being provided for the neighbouring islands of Dominica and St Vincent, and also for use in Paraguay."

Two St Lucia Amazons are depicted on the coat of arms – one on either side of the central shield. Each is perched on the motto: The Land, The Light, The People.

Paradise Park and the World Parrot Trust subsequently won *BBC Wildlife Magazine*'s Zoo Conservation Award for Excellence.

The St Lucia Amazon was first described in 1776. 'Deforestation by humans has been the most devastating factor for the parrots, followed by their capture for food and the wild bird trade and the effects of hurricanes,' says the DWCT.

'In three decades, starting with the 1950s, numbers had decreased from about 1,000 to a mere 100 and the parrot's habitat had shrunk to a fifth of what was already a tiny area. Although Jacquot has been officially protected since 1849, the legislation had been largely ignored.'

The St Lucia Amazon has graced a number of stamps, including a set of four WWF ones with values of 15 cents, 35 cents, 50 cents and 1 dollar issued by the island in 1987.

ST VINCENT

St Vincent Amazon

ST VINCENT PARROT

Amazon guildingii

The St Vincent Amazon is the national bird of St Vincent. An educational campaign undertaken in 1988 sought to educate St Vincent residents about deforestation and erosion by focusing on the threats to the St Vincent Amazon. The parrot's image was used on wide-ranging promotional material, including T-shirts, badges, posters, bumper stickers and billboards and in a Forestry Department slide/tape presentation. The St Vincent Amazon has also appeared on the island's stamps.

WHERE TO SEE One of the best places to see wild St Vincent Amazons is the St Vincent Amazon Reserve in the Upper Buccament Valley.

The species also occurs in the following IBAs: Colonarie Forest Reserve, Cumberland Forest Reserve, Dalaway Forest Reserve, Kingstown Forest Reserve, La Soufrière National Park, Mount Pleasant Forest Reserve and Richmond Forest Reserve. Captive-bred St Vincent Amazons are kept at St Vincent Botanical Gardens, Kingstown.

VULNERABLE

IUCN: Declined throughout most of the 20th century mainly as a result of hunting for food, trapping for the pet trade and habitat loss. Numbers started to recover in the early 1980s following conservation action. It is estimated that there are now around 800 mature individuals.

Size: Length about 40 cm (15.7 in). Adults weigh 660–700 kg (1,450–1,550 lb).

Description: A multihued parrot with highly variable plumage. There are two colour morphs – yellow-brown and green. The yellow-brown morph, which is the more common of the two, has a white head, blue-grey nape, bronze back and breast, black primary feathers with yellow bases, dark blue secondary feathers with orange bases, and a dark blue tail with a yellow terminal band.

Diet: Fruits, seeds and flowers. Mainly a canopy feeder but sometimes forages in partially cultivated areas.

Reproduction: Breeding takes place from January to June, peaks from February to May and occurs in small, loose groups, each pair defending its own nest site. Birds may not breed during particularly wet years. Two eggs are laid, usually in a cavity in a large, mature tree.

Range: Found only on St Vincent in the Lesser Antilles.

Habitat: Moist forests – mainly at an altitude of 125 to 1,000 metres (400–3,300 ft).

Cultural presence

Awareness of the St Vincent Amazon as St Vincent's national bird was boosted by more than 25 per cent as a result of the 1988 educational campaign. Puppet shows, primary and secondary school visits, two songs, a radio series on St Vincent's birds, an A–Z birdlife booklet, meetings with farmers, community groups and service clubs, newspaper articles, sermons and a monthly environmental newspaper called *Vincie's Nature Notes* were all used to deliver the conservation message, in addition to badges, posters, T-shirts and other such items.

First described and named by N. A. Vigors in 1836, the St Vincent Amazon has faced many threats over the years, the main ones being hunting for food, trapping for the pet bird trade and habitat loss.

BirdLife International says deforestation has been caused by a variety of things: forestry activities, banana cultivation expansion, charcoal production, the cutting down of nesting trees by trappers trying to reach young birds and natural catastrophes such as hurricanes and volcanic eruptions.

BirdLife adds that introduced Nine-banded Armadillos *Dasypus novemcinctus* have undermined large trees and led to them falling over, thus reducing the availability of suitable nesting sites.

'The genetic isolation of the separate sub-population may present further cause for concern.'

The Species Conservation Plan for the St Vincent Parrot prepared by St Vincent conservation biologist Lystra Culzac-Wilson and published by St Vincent's Ministry of Agriculture and Fisheries and the Loro Parque Fundacion in 2005 had three key objectives: improvement of the species' habitat and protection of wild birds until the natural population had reached its maximum sustainable level and the St Vincent Amazon could be removed from the IUCN's Red List; maintenance of a viable captive-breeding population within and beyond St Vincent; and the need to enshrine in national legislation and culture the protection and conservation of the St Vincent Amazon.

The St Vincent Amazon was just one of 16 bird species featured in a set of St Vincent stamps in 1970 on chalk-surfaced paper.

Secretarybird

Sagittarius serpentarius

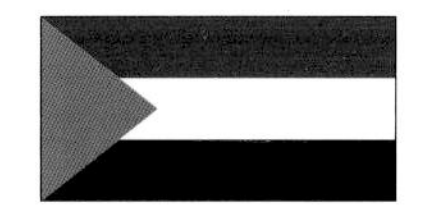

The Secretarybird, which replaced the rhinoceros as Sudan's national emblem during the government of President Jaffar Mohammed Nimeiri (1969–1985), symbolises might, sacrifice and dignity.

The Secretarybird forms the centrepiece of the Republic of Sudan's

Above: *Chicks are usually raised during the rainy season when food is plentiful.*

Left: *The Secretarybird preys on snakes, among many other animals, stamping on venomous adders, cobras and other species before swallowing them whole.*

VULNERABLE

IUCN: Evidence suggests that Secretarybird numbers are declining rapidly, probably as a result of habitat degradation, disturbance, hunting and trapping for trade. It is described as common to rare and local. The overall population is not thought to exceed five figures.

Size: Length 125–150 cm (49.2–59 in). About 1.2 m (3.9 ft) tall.

Description: The Secretarybird is unmistakeable and unlikely to be confused with any other species. It has long pink legs, an orange, raptor-like face, a hooked blue-grey bill, a black crest and a long tail. Its upperparts are grey, its underparts white and its flight feathers black. Sexes are identical.

Diet: Especially likes grasshoppers, locusts and beetles, but also eats many other small creatures, including snakes, lizards, frogs, mongooses, meerkats and other birds. Tends to dispatch its prey by kicking it and stamping on it before swallowing it whole. Uses its powerful feet to hold down larger animals while tearing them apart.

Reproduction: Usually two eggs are laid in a large platform-type nest at the top of a low tree – often a thorny *Acacia*.

Range: Sub-Saharan Africa. Native to 36 countries/territories, including Sudan and the newly created nation of South Sudan.

Habitat: Mainly open grassland with scattered trees.

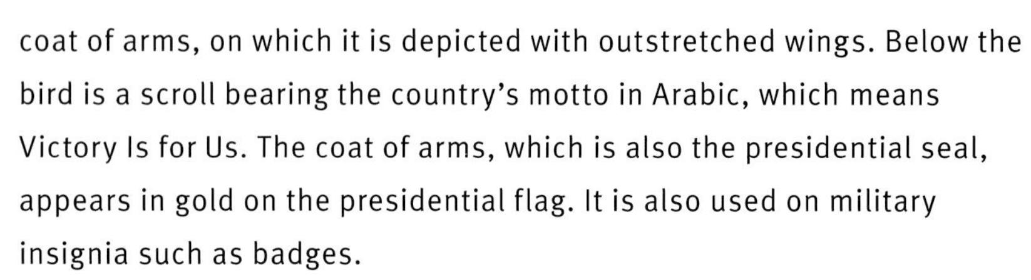

coat of arms, on which it is depicted with outstretched wings. Below the bird is a scroll bearing the country's motto in Arabic, which means Victory Is for Us. The coat of arms, which is also the presidential seal, appears in gold on the presidential flag. It is also used on military insignia such as badges.

Left : *The Secretarybird's 'spiky' loose crest is one of its most distinctive features.*

Above: *Although the Secretarybird spends most of its time on the ground, it is a capable flier, soaring on long, broad wings.*

WHERE TO SEE The species can potentially be seen wherever there is suitable habitat from very dry steppes to high, moist grasslands. Secretarybirds are most often found, however, in thorn-bush-dotted open grassland rich in the small animals on which this species preys. In Sudan it occurs in the Kordofan, Darfur and Blue Nile areas, and in the dry southern parts of the country.

Cultural presence

The Secretarybird's ability to kill venomous snakes and other animals regarded as vermin has resulted in it being admired in many quarters in Africa (it also appears on the coat of arms of South Africa). The origin of its strange name may have something to do with the fact that its eye-catching crest feathers are not unlike the quills used for writing in medieval times.

The Secretarybird appeared on one of the stamps in the definitive set issued by Sudan in 1991, and formed the watermark for the 10- and 20-Sudanese pound banknotes issued in July 2006.

The Secretarybird is an integral feature of Sudan's coat of arms.

SWAZILAND

Purple-crested Turaco

Tauraco porphyreolophus

The Purple-crested Turaco is the national bird of the Kingdom of Swaziland. Locally the bird is known as the *Igalagala* in the Saswati language. The flight feathers are significant in the ceremonial attire of the Swazi royal family.

WHERE TO SEE The species can be seen wherever there is suitable habitat – mainly moist woodland, riverside vegetation and evergreen thickets but also coastal forests, parks, gardens and exotic plantations. This frugivorous species often feeds at bird tables in southern Africa, eating guava, mulberries, paw-paw and maize-meal.

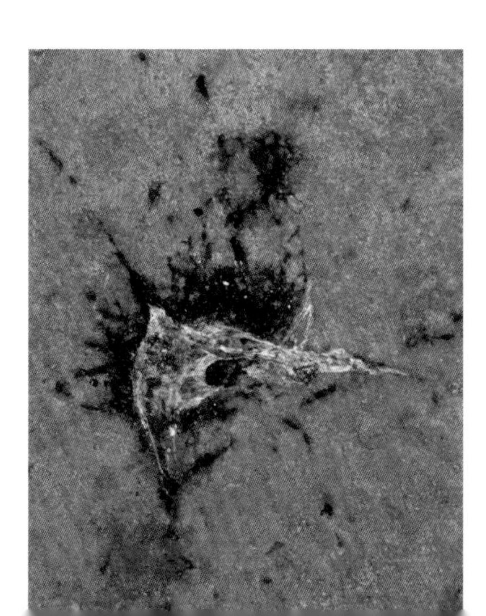

LEAST CONCERN

IUCN: Although the total number of mature individuals is unknown, this species is thought to be quite common in most parts of its southern African range. The population overall, however, is thought to be declining.

Size: Length 42–46 cm (16.5–18.1 in).

Description: Like many other turacos, the Purple-crested is a beautiful multicoloured bird – essentially a kaleidoscope of green, blue, purple and pink, with grey, black, brown and red also thrown in for good measure. There are two subspecies of this turaco. The Purple-crested Turaco subspecies found in Swaziland is the nominate race, *Tauraco porphyreolophus porphyreolophus*.

Diet: Fruits.

Reproduction: Two or three eggs are laid on a flimsy, well-concealed platform of twigs in a tree.

Range: Large. Native to 11 African countries: Burundi, Kenya, Malawi, Mozambique, Rwanda, South Africa, Swaziland, Tanzania, Uganda, Zambia and Zimbabwe.

Habitat: Damp woods, shrubby areas, coastal forests and plantations, as well as parks and gardens.

A dropping of the Purple-crested Turaco reveals seeds. This bird likes to feed on fruit such as guava and paw-paw.

The Purple-crested is one of 23 turaco species. Commonly called louries, they are endemic to sub-Saharan Africa.

Cultural presence

Top: *Turaco feathers are often used in Swazi royal headwear.*

Above: *The Purple-crested Turaco was depicted on this 1995 Swaziland stamp – part of a set of four turaco stamps.*

Turaco feathers are highly valued in Swazi royal circles. The red flight feathers of both the Purple-crested Turaco and Transvaal Turaco (*T. corythaix phoebus*) are worn by King Mswati III and those closest to him to show status, according to Clive Humphreys, a committee member of the International Turaco Society. Headdresses worn at the famous Ceremony of the Reeds and other important royal functions often incorporate the crimson primary and secondary flight feathers of local turaco species. The king and his many wives keep a large number of turaco feathers with which to adorn their hair. One of King Mswati III's young wives wore turaco feathers in her headdress and hair while attending a crowned heads dinner in Windsor, England, during Queen Elizabeth II's diamond jubilee celebrations in 2012.

Swaziland Tourism (UK) says red wing feathers symbolise intelligence and royalty. The king wears three red feathers as a crown, princes wear up to three (normally on the side of the head) and princesses only wear them during the annual Umhlanga and Incwala ceremonies. It is a symbol of excellence and the king awards outstanding members of his regiments with a red feather for loyal service. It is also a royal welcome that Swazis extend to their visitors.

The Purple-crested Turaco has appeared on a number of Swaziland stamps and was one of four turaco species illustrated in a set issued in 1995. A turaco feather is incorporated in the logo of the Swaziland Tourist Authority.

Eurasian Blackbird

Turdus merula

The Eurasian Blackbird is the national bird of Sweden. It was chosen as the country's national bird in 1962 following a newspaper poll.

WHERE TO SEE This adaptable species can be seen in most habitats.

Cultural presence

A member of the thrush family (Turidae), the Eurasian Blackbird is known in Sweden as the *Koltrast*. Although it is not clear precisely why the Eurasian Blackbird became Sweden's avian icon, it is worth pointing out that this species is noted for its mellifluous song. The Eurasian Blackbird was featured on a 1970 Swedish Christmas stamp.

A Eurasian Blackbird appeared on this 1970 Swedish stamp.

***Source:** © Sweden Post Stamps*

LEAST CONCERN

IUCN: It is estimated that there are 120 to 246 million Eurasian Blackbirds in Europe – 50–74 per cent of the global total.

Size: Length 24–27 cm (9.5–10.6 in).

Description: The aptly named Eurasian Blackbird is just that – a black bird with a bright orange-yellow bill and matching eye-ring (as far as the male of the species is concerned). Females are dark brown and paler. There are various subspecies of Eurasian Blackbird. The one occurring in Sweden is the nominate race of *Turdus merula merula*.

Food: Omnivorous. Feeds largely on worms, insects, larvae and other invertebrates, but quite happily takes a range of fruits, including berries, and even garden scraps.

Reproduction: Usually three or four eggs are laid in a cup-like nest made of grass and twigs concealed in a bush or tree. Predation by cats and crows can be high in urban and suburban areas.

Range: Huge. Found in many parts of the world.

Habitat: Originally a forest and woodland bird, the Eurasian Blackbird is now found in a wide range of environments, from farmland and open countryside to city centres and gardens.

Juvenile Eurasian Blackbirds, such as this one, are brown and spotted.

An adult male Eurasian Blackbird at its nest where there are lots of hungry mouths waiting to be fed.

Taiwan Blue Magpie

FORMOSAN MAGPIE

Urocissa caerulea

LEAST CONCERN

IUCN: Although precise numbers have yet to be quantified, this species is not thought to be uncommon.

Size: Length 63–68 cm (24.8–26.8 in).

Description: Striking bird with a black head, nape and upper breast, a dark blue-purple body, a very long tail, yellow eyes and a red bill and legs. There are white markings on both the wings and the tail. Sexes are similar.

Diet: Omnivorous – from invertebrates to berries and figs.

Reproduction: Five or six eggs are laid in a large platform-type nest in a tree. This species is Taiwan's only known bird to breed cooperatively.

Range: Endemic to Taiwan.

Habitat: Dense, broadleaved mountain forests.

The Taiwan Blue Magpie was voted national bird in 2007. It has appeared on various Taiwan stamps and seems likely to enjoy an even higher profile following its adoption as this island nation's avian icon.

WHERE TO SEE The Taiwan Blue Magpie is not an easy bird to observe well, given that it is elusive and wary, and frequents thick forests.

Cultural presence

Also known as the 'long-tailed mountain lady', this species was chosen as Taiwan's unofficial national bird in 2007 following an international poll organised by the Taiwan International Birding Association. More than one million votes were cast by people in 53 countries (987,063 via computer and 70,000 in the form of paper ballots), with the Formosan Magpie winning by a handsome margin with 491,572 votes. The runner-up, with 277,178 votes, was the Mikado Pheasant *Syrmaticus mikado.*

The Formosan Magpie is featured in the logo of the Taiwan International Birding Association (TIBA).

The long, graduated tail of the Taiwan Blue Magpie can be clearly seen in this excellent flight photo.

THAILAND

Siamese Fireback

Lophur diardi

Also called Diard's Fireback after French naturalist Pierre-Medard Diard, the Siamese Fireback is one of 49 known pheasant species. Although it was officially adopted as the national bird of Thailand in 1985, it does not appear to have an especially high profile within the country. In fact, like other pheasants in the *Lophura* genus, very little is known about the ecology and behaviour of this bird in the wild.

Within its range as a whole, the Siamese Fireback is threatened not only by extensive deforestation, but also by hunting and snaring. It seems that hunters often eat Siamese Firebacks while searching for higher value species.

LEAST CONCERN SINCE 2011

IUCN: Formerly Near Threatened. Moved down a category because the rate of decline of this species is now suspected to be slow to moderate and not as rapid as once believed. Overall population estimated to be up to 50,000. Around 5,000 individuals are estimated to live in Thailand where it lives mainly in the north-east and south-east of the country.

Size: Length 60–80 cm (23.6–31.5 in). Males are around 20 cm (8 in) longer than females and weigh much more.

Description: Males have bright red facial wattles, a dark, plume-like crest, a mostly grey, finely vermiculated body but with a bright yellow back patch, a long, dark, curved glossy tail and red legs. Females, which lack the crest, have a two-tone appearance – large areas of white-barred black above and predominantly brown below.

Diet: Thought to be omnivorous, eating fruits including berries, and insects, worms and such like.

Reproduction: Poorly known. The species is believed to lay four to eight eggs in a nest on the ground.

Range: Native to Cambodia, Laos, Myanmar (Burma), Thailand and Vietnam. There are estimated to be around 5,000 individuals in Thailand, where this species is an uncommon to locally common resident. It is found in northern, north-eastern and eastern Thailand.

Habitat: Both primary and secondary lowland forests.

WHERE TO SEE Among the places in Thailand where the Siamese Fireback can be seen are Khao Yai National Park, Phu Khieo Wildlife Sanctuary and Sakerat Biosphere Reserve. Researchers had 232 encounters with Siamese Firebacks during 136 days of fieldwork in Khao Yai between May 2002 and April 2003. Footage of a small flock of Siamese Firebacks in Sakerat has been uploaded to YouTube.

The Siamese Fireback occurs in a range of habitats from dense evergreen and bamboo forests to scrub and alongside forest tracks and roads, which is fortunate in light of the deforestation in Thailand and, in turn, its habitat loss.

The species seems to have a high tolerance to habitat change and disturbance. It can also survive at high altitudes; it has been known to occur very occasionally up to 1,300 m (4,250 ft) above sea level. It usually resides below 500 m (1,600 ft) above sea level.

Top: *An adult female Siamese Fireback is rufous brown with black-and-white banded wings and tail. Females lack the male's head tuft.*

Middle: *The male Siamese Fireback has distinct yellow plumage in the middle of its back, to which its name 'fireback' refers.*

Above: *Siamese Firebacks usually lay four to eight pale, round eggs.*

Cultural presence

The image of the Siamese Fireback is not widely used in Thailand, despite being the country's official avian emblem.

However, the Siamese Fireback was one of eight species illustrated in a 1967 set of Thai bird stamps. An image of two Siamese Firebacks moving through native Thai vegetation appeared on the reverse of a WWF 25th-anniversary silver commemorative coin in 1987.

The Siamese Fireback is a favourite with birdwatchers visiting Thailand. The Khao Yai National Park has well-developed trails for birding tourists; its mixture of grassland and forest ensures a variety of bird species can be seen. There are over 50 km (30 miles) of hiking trails to be explored and there are two major entrances, one to the north in the Nakhon Ratchisma Province and the other south in the Prachinburi Province.

A Siamese Fireback appeared on one of eight Thai bird stamps in 1967.

TRINIDAD AND TOBAGO

Scarlet Ibis (TRINIDAD)

Eudocimus ruber

Rufous-vented Chachalaca (TOBAGO)

Ortalis ruficauda

The Republic of Trinidad and Tobago is unique in having not one, but two, official national birds, the Scarlet Ibis for the island of Trinidad and the Rufous-vented Chachalaca for the island of Tobago.

Both the Scarlet Ibis and the Rufous-vented Chachalaca are depicted on the national coat of arms of Trinidad and Tobago, with the former standing to the left of the central shield, and the latter to the right. The two birds are shown with their wings open. A third bird, a hummingbird, is also featured. The coat of arms dates back to 1962 when the twin-island nation achieved independence from the UK.

The Scarlet Ibis has appeared on many Trinidad and Tobago stamps, including a set of five different 50-cent ibis stamps issued in 1980 and a set of four different WWF ibis stamps, each with a face value of 5.25 dollars, released in 1990. It has also been depicted on Trinidad and Tobago coins, usually as part of the coat of arms, but also in its own right on a 5-dollar coin.

The Rufous-vented Chachalaca appeared on a stamp in 1969.

SCARLET IBIS

LEAST CONCERN

IUCN: Common and widespread within its range. Overall population very large, although precise numbers are unknown. It is thought that this species is declining as a result of habitat loss.

Size: Length 55–63 cm (21.7–24.8 in).

Description: Unmistakeable on account of adults' vivid scarlet plumage. Black wingtips and pinkish legs. Strongly decurved bill. Closely related to the American White Ibis *E. albus*, with which it hybridises in certain parts of its range. In fact, some sources consider the two to be the same species, with *ruber* being a colour variant of *albus*.

Diet: Mainly crustaceans. Especially fond of fiddler and other crabs. Also eats snails, bivalves and other molluscs, as well as insects and fish.

Reproduction: Usually two eggs are laid in a small platform-type stick nest in a tree, often a mangrove.

Range: Very large. Native to 11 countries/territories, including Trinidad and Tobago, and a vagrant in nine others.

Habitat: Coastal mangrove swamps, estuaries, mudflats and wet inland savannahs.

Scarlet Ibises are stunning birds; they were once hunted for their feathers.

Compared to the Scarlet Ibis, the turkey-like Rufous-vented Chachalaca is a rather drab bird. But what it lacks in looks it makes up for in noise.

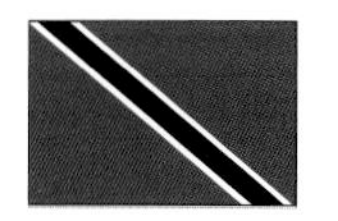

RUFOUS-VENTED CHACHALACA

LEAST CONCERN

IUCN: Global population has yet to be evaluated, but is thought to be stable.

Size: Length 53–61 cm (20.9–24 in).

Description: Turkey-like bird with a small head, a long, strong neck, short wings and a long, broad tail. Fairly nondescript in terms of colour – adults are largely buff, grey-brown and olive-brown. Grey bill, legs and crown, and bare red throat. A very noisy species, especially at dawn. There are two subspecies, *O. r. ruficauda* being the one found on Tobago.

Diet: Fruits, shoots, leaves and flowers.

Reproduction: Three to four eggs are laid in a twig-and-leaf nest in a tree.

Range: Native to five countries/ territories. Common and widespread on Tobago. Not found on Trinidad.

Habitat: Forests, brush/scrub, abandoned agricultural land and gardens.

WHERE TO SEE The best place to see the **Scarlet Ibis** on Trinidad is at the Caroni Bird Sanctuary – the species' major roosting and breeding site. Although 2,500 nests were counted in June 1963, the Scarlet Ibis abandoned Caroni as a breeding colony after 1970 and occurred there only in the non-breeding season, when up to 15,000 birds were present. Nowadays, Scarlet Ibis are once again breeding in good numbers at Caroni – a 5,611-hectare (13,865-acre) area of mangroves, marshes and tidal mudflats. The Caroni Bird Sanctuary is a major attraction for birdwatchers. Six different tours are available, including a dedicated Scarlet Ibis photography tour.

The **Rufous-vented Chachalaca** can be seen, for example, at the Main Ridge on Tobago and on Little Tobago Island. The Main Ridge, which forms the spine of the north-eastern part of Tobago, provides habitat for four restricted-range bird species, including the Rufous-vented Chachalaca, which was said to be abundant in the area in 2006. Little Tobago – a small island about 2 km (1.2 miles) off the north-east coast of Tobago itself – supports a small population of the birds.

Cultural presence

The Scarlet Ibis is a high-profile bird in Trinidad and Tobago, not least because of its popularity with birdwatchers and general tourists, who make tracks for Caroni Bird Sanctuary to see the guaranteed spectacle of flocks of bright pink birds flying in to roost at night. Local ornithologist Winston Nanan has described the roost as resembling 'living Christmas trees'.

Scarlet Ibises were once hunted in Trinidad for their feathers, which were highly prized for hats and carnival costumes. Despite the fact that this species has long been Trinidad's national bird and that firearms are banned in Caroni, some poaching still takes place.

Known locally as the Cocrico, the **Rufous-vented Chachalaca** is regarded as something of a pest by farmers on Tobago because it is present in large numbers and is said to damage and destroy crops.

The Rufous-vented Chachalaca appeared on the obverse of Trinidad and Tobago's $10 banknotes.

Grey Crowned Crane

Balearica regulorum

Known throughout Uganda, where it is the national bird, the Grey Crowned Crane is considered by many people to be sacred and an indicator of wetland health.

The *gibbericeps* subspecies is featured prominently on Uganda's coat of arms. It stands to the right of a shield and spears facing the Ugandan Kob *Kobus kob thomasi*. A stylised image of the bird also appears on the national flag.

WHERE TO SEE South-western Uganda supports the largest number of Grey Crowned Crane breeding pairs in the country. Local hotspots include the Bushenyi, Kabale and Ntugamo Districts.

The Grey Crowned Crane is a beautiful bird, its most eye-catching feature being the golden crown.

UPLISTED FROM VULNERABLE TO ENDANGERED IN 2012

IUCN: Grey Crowned Crane numbers have declined rapidly over a period of 45 years as a result of habitat loss, and the illegal taking of birds and eggs from the wild for food, traditional use, domestication or sale on the international market. Flying into power lines and electrocution by them also results in many Grey Crowned Crane deaths in South Africa and, increasingly, Uganda. It is the *gibbericeps* subspecies that is found in Uganda, of which there are less than 10,000 individuals (possibly fewer than 5,000).

Size: Length 100–110 cm (39.4–43.3 in).

Description: Striking golden-yellow crown, mainly grey body, largely white wings, and black legs and toes. *Gibbericeps* has more red on its cheeks than the nominate race *regulorum*. Sexes are almost identical, although male birds are usually slightly bigger than females.

Diet: Omnivorous. Seed heads and grass tips; groundnuts, soybeans, maize and millet in agricultural areas; insects and other invertebrates; small vertebrates such as frogs, lizards and crabs.

Reproduction: Two to four eggs are laid on a platform of flattened grasses and sedges. Nests are invariably in or adjacent to wetlands and concealed by surrounding vegetation. Grey Crowned Cranes have the biggest average clutch size of any crane species.

Range: Very large. Native to 17 countries in eastern and southern Africa. Although the Grey Crowned Crane is non-migratory, it does undertake local and seasonal movements.

Habitat: A mixture of wetlands, grasslands and cultivated areas.

Far left: *The Grey Crowned Crane eats all manner of things, from seeds to small vertebrates.*

Left: *All cranes generally pair for life. Up to 60 Grey Crowned Cranes have been observed simultaneously performing their breeding displays.*

Cultural presence

The International Crane Foundation has teamed up with local conservation bodies in Uganda, such as Nature Uganda, to encourage and support the creation of community-based conservation efforts in the Lake Victoria Basin. Here the remaining threatened wetlands are vital not only to cranes and other wildlife but also to local communities, providing clean drinking water, supporting lucrative fisheries and yielding a wealth of natural resources such as papyrus and reeds for baskets and other local products. Lake Victoria is the world's third largest lake and nearly half of the 50 million people who live in Uganda and Kenya dwell within the greater basin.

Uganda-born Jimmy Muheebwe-Muhoozi, formerly a tutor at a primary teacher training college in the Bushenyi District of south-west Uganda, and now working for Nature Uganda, formed a number of school wildlife clubs and revitalised dormant ones to raise awareness about the plight of cranes and the wetlands on which they depend.

South-west Uganda is densely populated, most of the wetlands having been developed for farming and hillsides having been cleared for the cultivation of potatoes and other crops. Muheebwe-Muhoozi realised that the future of cranes and other wildlife depended on establishing a new relationship between Ugandans and their wetlands.

Today, there is much greater awareness in south-west Uganda of the need to value and conserve cranes and associated wetlands. During the past decade or so the wide-ranging educational and conservation activities in Uganda have included using music, dance and drama to draw attention to the plight of cranes, monitoring the birds, encouraging local people not to drain and cultivate swamps, planting indigenous trees adjacent to wetland areas to replace those felled for firewood or farming, and using bee-keeping/honey harvesting and organic farming as ways of helping to alleviate poverty.

The Grey Crowned Crane has adorned a number of Ugandan stamps over the years and was one of eight bird species depicted in a set of bird stamps issued in 1987.

The Grey Crowned Crane was one of 14 bird species featured on a set of Ugandan stamps in 1965.

An athlete carrying Uganda's flag, at the centre of which is a stylised image of a Grey Crowned Crane.

European Robin

Erithacus rubecula

The European Robin has never been officially adopted as the UK's national bird. It was, however, declared the country's unofficial avian emblem in the early 1960s after a survey run by *The Times* newspaper. Its image has probably been employed more often and in a greater variety of ways in UK society than that of any other bird. Images abound on Christmas cards, stamps, calendars, coasters, mugs, dishes, embroidered badges, pin badges (including a limited edition badge produced specially for the London 2012 Olympic Games) and tea towels.

WHERE TO SEE The European Robin is a common and familiar bird that is nowadays just as likely to be seen in towns and cities as in the wider countryside.

LEAST CONCERN

IUCN: European breeding population is estimated to be between 43 and 83 million pairs.

Size: Length 14 cm (5.5 in).

Description: Adults are an eye-catching combination of a diagnostic orange-red face and breast, an off-white belly, buff-brown flanks, a brown back and tail and dark eyes.

Diet: Invertebrates, fruits and seeds. Very partial to both live and dried mealworms at garden feeding stations.

Reproduction: Four to six eggs are usually laid. Eclectic choice of natural and man-made nest sites, from tree hollows, tree roots and rock crevices to nestboxes, old kettles, watering cans, flowerpots and other suitable containers.

Range: Extremely large – the UK, continental Europe, parts of Asia and North Africa. Europe accounts for 75–94 per cent of this species' global range. UK birds are mainly resident and joined in winter by birds from Europe.

Habitat: From gardens, parks and hedgerows to woodland, churchyards, allotments and the like. Before European Robins colonised gardens they were essentially woodland birds, feeding on insects and other invertebrates displaced by wild cattle, Wild Boar *Sus scrofa* and deer.

Far left: *Robins nest in all manner of places, including old kettles and watering cans.*

Left: *This is probably what Robins are seen doing in gardens more than anything else – plucking a juicy earthworm from freshly dug soil.*

Robins often appear fairly relaxed and comfortable in human company, especially around gardeners. They have a reputation for being friendly. However in the bird world they are aggressive defenders of their territory.

Cultural presence

The European Robin is deeply rooted in folklore. It is said, for example, that a Robin plucked a thorn from the Crown of Thorns worn by Jesus before and during his crucifixion. A drop of Christ's blood supposedly fell onto the Robin, turning its breast red – hence the alternative name for the bird of Robin Redbreast. Another myth is that the bird's breast was scorched red while the bird conveyed water to souls in purgatory.

Robins cover the bodies of two children in the tale of Babes in the Wood.

During William Shakespeare's time (1564–1616), and possibly much earlier, the species was associated with charity and piety. Its popularity soared to new heights in the Victorian era because the first postmen wore red uniforms and were nicknamed Robins, while the bird itself was often shown on Victorian Christmas cards carrying a card or letter like real postmen.

Christmas simply would not be Christmas now without the Robin, and designers and manufacturers of Yuletide cards would certainly be lost without this much-loved bird. It is depicted photographically or illustratively on Christmas cards in a seemingly never-ending range of poses, from being surrounded by red-berried Holly to perching on the handle of a garden fork or spade or sitting on a snow-covered twig.

Robins are virtually synonymous with frost, snow and winter in the UK. This government poster campaign ran between 1942 and 1947, featuring a Robin. It encouraged people to save fuel resources.

The European Robin has also been immortalised in song, verse and literature. 'When the Red, Red Robin (Comes Bob, Bob, Bobbin' Along)' – penned by Harry M. Woods in 1926 – was a hit for several singers, including Al Jolson and Doris Day.

Robins cover the bodies of two dead children with leaves in the well-known children's story *Babes in the Wood*. Poet William Blake (1757–1827) famously opined that 'a robin redbreast in a cage puts all heaven in a rage'. According to one old wives' tale, if a Robin was the first bird seen by a maiden on Valentine's Day (14 February), this meant that she would marry a sailor or some other nautical man.

It seems no one knows who wrote the centuries-old nursery rhyme, 'Who Killed Cock Robin?' in which Cock Robin is killed by a Sparrow with a bow and arrow. 'All the birds of the air fell a-sighing and a-sobbing when they heard the bell toll for poor Cock Robin', concludes the rhyme. Interestingly, a stained glass window

Cock Robin was killed by an arrow in the timeless nursery rhyme Who Killed Cock Robin?

at Buckland Rectory (*c.* 1480, later altered) near Cirencester, Gloucestershire, depicts a Robin being killed in this very manner. The Robin fares better in another rhyme, 'Little Robin Redbreast', for it escapes the clutches of a cat, challenging the latter to 'Catch me if you can'.

If a Robin enters a house on Dartmoor, Devon, this was and still is considered to be a bad omen (usually of death), especially if the bird utters a 'teep, teep' call, according to Tim Sandells on his Legendary Dartmoor website. Furthermore, any Christmas cards portraying a Robin were once considered bad luck. Tim, who used to live on Dartmoor and has spent many hours walking on, reading about and conducting research in respect of the area, says: 'I can remember granny always saying that if anybody sent a card with a Robin they were no friend as they were wishing you bad luck. She would immediately tear up any such card.' He added that his grandmother sent a few Robin cards to people she disliked.

Tim Sandells goes on to say: 'If anybody dared to steal an egg from a Robin's nest, the penalty would be that all the milk produced in the area would be discoloured.

'If a Robin was ever seen to hop over the threshold of a door, then it was an omen of forthcoming debt.

'On seeing the first Robin of the New Year, it was said that any wish made would come true. If, however, the bird flew away before the wish was finished, this would end up with 12 months' bad luck.

'The Robin has also been used to forecast the weather. If it is heard singing from cover, then bad weather is on the way. Conversely, if the Robin sings from an open branch, then good weather is coming.

'I have also heard tell that if a Robin ever finds a dead body, it will cover the corpse's face with moss and leaves as a mark of its love for the human race. This tradition may well come from the story of *Babes in the Wood*, as it was a Robin that covered the faces of the dead children. Although I know of no example on Dartmoor, you can soemtimes see a Robin carved on gravestones, possibly denoting that the deceased was a friendly person.'

No fewer than five English football clubs are nicknamed Robins: Altrincham, Bristol City, Charlton Athletic, Cheltenham Town and Swindon Town. The colour of each club's home kit is the same as that of a Robin's breast.

The Robin has appeared on more Christmas cards and Christmas stamps in the UK than any other bird species.

Bananaquit

Coereba flaveola

The Bananaquit was named official national bird in 1970. It is depicted on the US Virgin Islands' coat of arms perched on a Yellow Trumpetbush *Tecoma stans*, and surrounded by the outlines of the islands of St Thomas, St John and St Croix.

WHERE TO SEE Flower-rich gardens are as good a place as any to look for the ubiquitous Bananaquit. It is also found in Virgin Islands National Park on St John, among other places.

As many islanders know, Bananaquits love not only hummingbird feeders containing sugared water but also bowls of sugar. Not for nothing are they often referred to as Sugar Birds. Bananaquits in the US Virgin Islands often boldly enter people's homes to take sugar from a table in full view of diners.

Bananaquits commonly visit gardens, flitting from one plant to another in search of energy-giving nectar. These energetic little birds can also be found in a wide range of other environments in the US Virgin Islands including shrub-covered areas, mangroves and some woodland. Bananaquits can be seen, too, in Virgin Islands National Park – along with many other birds, including hummingbirds and wintering US warblers.

LEAST CONCERN

IUCN: The total population is estimated to be anywhere between five and 50 million individuals and is believed to be stable.

Size: Length 10.5–11 cm (4.1–4.3 in).

Description: An eye-catching species with bright yellow underparts, contrasting dark upperparts, a prominent white eye-stripe and a short, decurved bill. Sexes are similar. There is considerable geographical variation within the species and there are more than 40 subspecies. The race *Coereba flaveola sanctithomae* is the one found in the US Virgin Islands.

Diet: Mainly nectar, but also some fruits including berries, as well as insects.

Reproduction: Two to four eggs are laid in a spherical nest entered via a side entrance.

Range: Very large. Native to nearly 50 countries/territories, including the US Virgin Islands.

Habitat: Wide-ranging – from shrubs, forests and mangrove areas to parks and gardens.

Above: *Like hummingbirds, Bananaquits are active feeders, but these two have been photographed at rest, enabling us to see well their two-tone plumage and prominent white eyebrows.*

Left: *The Bananaquit feeds mainly on nectar which it obtains either by inserting its decurved bill directly into flowers or by piercing the base of them.*

Cultural presence

The Bananaquit is also known as the Yellow Bird or Yellow Breast and Sugar Bird, the former on account of its yellow underparts and the latter because it loves sugar. This familiar species commonly frequents gardens where nature-loving residents put out bowls of granular sugar, or feeders of sugar water, and can become very tame as a result.

The Bananaquit image appears on the US Virgin Islands' government seal and it also appears on the reverse of the quarter coin.

The seal of the government of the US Virgin Islands showing a Bananaquit on a Yellow Trumpetbush Tecoma stans.

UNITED STATES OF AMERICA

Bald Eagle

Haliaeetus leucocephalus

Long synonymous with liberty, courage and strength, the Bald Eagle has been the USA's national bird since 1782, when it was incorporated into the country's Great Seal. The bird was and still is depicted on the Great Seal holding an olive branch in its right talon as a symbol of peace and a bundle of 13 arrows in its left talon as a symbol of war.

Over the years, this familiar avian icon has appeared officially and unofficially on all manner of objects, from the Seal of the President of the USA, the Presidential Flag, the Mace of the House of Representatives, the Congressional Medal of Honour and USAF, the Navy, Marine Corps and American Airlines emblems to various US coins, one-dollar bills and stamps. In fact, the Bald Eagle is almost certainly the world's best-known and most widely used national bird.

WHERE TO SEE One of the best places to see large numbers of these iconic birds is the 19,425-hectare (48,000-acre) Chilkat Bald Eagle Preserve near Haines, Alaska, which was created in 1982 by the State of Alaska to 'protect and perpetuate the world's largest concentration of Bald Eagles and their critical habitat' and to sustain and protect the associated salmon runs. More than 3,000 eagles have been recorded at Chilkat Preserve from October through February during the Fall

LEAST CONCERN

IUCN: Bald Eagle numbers plunged during the 20th century, mainly as a result of pesticide poisoning which rendered birds sterile or resulted in them laying brittle eggs; both legal and illegal shooting (bounties used to be paid for killing eagles in Alaska); and habitat loss. Following the 1940 Bald Eagle Protection Act and other measures, the population has bounced back to 300,000 mature individuals.

Size: Length 76–91 cm (30–35.8 in); huge wingspan of 1.68–2.44 m (5.5–8 ft). Females are bigger than males.

Description: Seriously impressive raptor. Both adult males and females have a dark brown body, a contrasting white head and tail and a large yellow bill. There are two subspecies, both of which are found in the USA – the nominate race *Haliaeetus leucocephalus* occurring in southern parts of the country and *H. l. washingtoniensis* in northern areas.

Diet: Although the Bald Eagle feeds mainly on fish, this powerful, opportunistic predator also quite happily takes everything from mammals and birds to reptiles and invertebrates, as well as carrion and human flotsam and jetsam.

Reproduction: A large and bulky, vegetation-lined nest of sticks is usually built in the main fork of a tree up to 60 m (197 ft) tall. The nest, which can be up to 4 m (13 ft) in diameter and 2.5 m (8 ft) deep, is often in a conifer. Two or three eggs are generally laid. Birds mate for life.

Made of sticks, Bald Eagle nests are so large they can easily be seen from a distance.

Range: The Bald Eagle is a widely distributed bird of prey breeding in the USA, Canada and Mexico and on Saint-Pierre and Miquelon Islands off the Newfoundland coast. It is common to very local and scarce.

Habitat: Breeds on or near lakes, rivers and estuaries, as well as in coastal areas. Often found in similar habitats in winter, but also in a wide range of other environments, including dams, rubbish dumps, sheltered valleys and arid terrain in western parts of America.

An awesome predator and a joy to behold both in flight and action, this Bald Eagle – talons outstretched – is about to snatch a fish from the water.

Congregation. According to the American Bald Eagle Foundation, more Bald Eagles occur in Alaska – an estimated 50,000 birds – than anywhere else in the USA.

Brackendale, where a world record 3,769 Bald Eagles wintered in 1994, is situated to the north of Vancouver, British Columbia, Canada, and is home to the Brackendale Winter Eagle Festival, which takes place throughout January. Volunteers have been participating in the annual Brackendale Winter Bald Eagle Festival and Count since 1986. More than 1,000 were counted on the 5th january, 2014.

Anyone wanting to get up close and personal with Bald Eagles should make tracks for the American Eagle Foundation's Pigeon Ford Eagle Center at Dollywood, Pigeon Forge, Tennessee, which boasts the world's biggest Bald Eagle breeding facility. Dozens of captive-bred birds from the centre have been released into the Great Smoky Mountains area of Tennessee and elsewhere.

Cultural presence

Although the Bald Eagle is synonymous with the USA and has been the country's national bird for well over 200 years, this iconic raptor nearly did not become North America's avian icon. One of the Bald Eagle's opponents was none other than Benjamin Franklin, who described this raptor in a letter to a friend as being of 'bad moral character … generally poor … and often very lousy'. Franklin thought that the Turkey *Meleagris gallopavo* was 'a much more respectable bird' and more worthy of the honour. Not everyone in Congress shared Franklin's views, however, and the Bald Eagle was ultimately the victor.

Bald and Golden Eagles *Aquila chrysaetos* have long been revered and held sacred in the traditions, culture and religion of Native Americans. Eagles, it is believed, are divine messengers entrusted with the task of carrying the prayers of mankind between the World of Earth, where people live, and the World of Spirit, where the Creator and grandfathers dwell.

Eagles and their feathers are accorded the greatest care and shown the deepest respect, according to the American Eagle Foundation. Native American warriors were only allowed to wear eagle feathers in their hair, headdresses or war bonnets if they had carried out a brave deed considered sufficiently worthy by their tribal council. To be given an eagle feather is the highest of all Native American honours – a mark of gratitude, love and respect.

Whistles made from eagle wing bones are used in the sacred sun dance ceremony of the Lakota and Dakota people. Eagles are featured heavily in the legends, stories and art of most, if not all, Native Americans.

The National Eagle and Wildlife Property Repository at Rocky Mountain Arsenal National Wildlife Refuge in Commerce City, Colorado, established by the US Fish and Wildlife Service (USFWS) in the early 1970s, stores the remains of dead Bald and Golden Eagles received by the USFWS so they can be legally used by enrolled members of federally recognised Native American tribes for religious purposes. Demand is high and supplies are limited. Orders for a complete carcass of an adult or immature Bald Eagle can take around two and a half years to fulfil, a pair of eagle wings approximately one year and 10 loose, quality feathers about six months. Requests for talons and heads are eventually met.

Indigenous American Indians are permitted to use eagle feathers in headdresses, such as this war bonnet.

The Great Seal of the USA on which the Bald Eagle is depicted holding an olive branch (peace symbol) and 13 arrows (war symbol).

The Bald Eagle's image has appeared on all manner of official and unofficial objects over the years, including stamps, coins and passports.

PASSPORT

United States of America

WIN THE WAR
3¢ 3¢
UNITED STATES POSTAGE

USA state bird emblems

Northern Flicker

California Quail

American Robin

ALABAMA

Northern Flicker *Colaptes auratus* A symbol of Alabama since the Civil War, the Northern Flicker or Yellowhammer Woodpecker was designated the state's official bird in 1927.

Nicknamed the Yellowhammer State, Alabama is the only US state to have chosen a woodpecker as its avian icon.

ALASKA

Willow Ptarmigan *Lagopus lagopus* This species – the biggest of three Arctic grouse found in Alaska – was chosen in 1955 by schoolchildren as a symbol of the 'Great Land' in preparation for statehood. It was officially adopted as the state bird when Alaska was declared the 49th state in 1960.

ARIZONA

Cactus Wren *Campylorhynchus brunneicapillus* The largest wren in the USA and a very common desert resident, the starling-sized Cactus Wren was chosen by the Arizona Legislature in 1931.

ARKANSAS

Northern Mockingbird *Mimus polyglottos* This species was adopted as the state's official avian icon on 5 March 1929 by the 47th General Assembly of Arkansas. The Mockingbird's image is used widely in educational materials for school history classes.

CALIFORNIA

California Quail *Callipepla californica* The Quail was adopted as California's state bird in 1931 following the selection of this bird as an appropriate state symbol by the Audubon Society. The relevant legislation, Assembly Bill No. 776, was signed by Governor James Rolph Jr on 12 June 1931 and took effect on 14 August of that year. It seems the Quail was chosen to represent California because it is native to America's west coast. It is also found in Idaho, Nevada, Utah and along Mexico's west coast.

Willow Ptarmigan

COLORADO

Lark Bunting *Calamospiza melancorys* The black-and-white migratory Lark Bunting was formally adopted as Colorado's state bird on 29 April 1931.

CONNECTICUT

American Robin *Turdus migratorius* This species was adopted as Connecticut's state bird in 1943 by the General Assembly. A migratory bird, the American Robin was first called a robin by early colonists in fond memory of the British robin, although the two species are completely different in both size and plumage.

DELAWARE

Blue Hen Chicken Delaware is one of only two US states to have a chicken as its avian icon, the other being Rhode Island. The precise origin of the name Blue Hen Chicken is unclear, although there are various theories.

Historian C. A. Weslager, for example, believes the name was given to the Delaware Regiment, founded in 1775, because of the smart uniform worn by its soldiers – white breeches, waistcoat, stockings, blue coats, black shoes and red feather plumes on peaked, leather hats resembling a chicken's comb.

Northern Mockingbird

Brown Thrasher

Mountain Bluebird

Dr Paul H Sammelwitz of the University of Delaware, writing in a paper entitled *The Delaware Hen: Fact and Fancy*, says Weslager suggested the regiment 'surely conjured up the image of gamecocks. Since many of the men were from Kent County, where cockfighting was popular, the analogy was natural.'

On 14 April 1939, Delaware General Assembly adopted the Blue Hen Chicken as its state bird.

FLORIDA

Northern Mockingbird *Mimus polyglottos*

Florida adopted the Mockingbird as its state bird on 23 April 1927. Senate Concurrent Resolution No 3 declared that the 'melody of its music has delighted the ear of residents and visitors to Florida from the days of the rugged pioneer to the present comer.' The Resolution added that this species was of 'matchless charm' and found throughout the state.

GEORGIA

Brown Thrasher *Toxostoma rufum*

According to the April 1932 edition of *Nature* Magazine, it was the schoolchildren of Georgia in 1928 who first selected the Brown Thrasher as the state's avian icon, although no legislative action was taken at that time. Several years later, on 6 April 1935, Governor Eugene Talmadge also declared the Brown Thrasher to be Georgia's state bird. It wasn't until 20 March 1970, however, following pressure from the Garden Clubs of Georgia, that the General Assembly passed Joint Resolution No 128 declaring the Brown Thrasher to be the official state bird and the Bobwhite Quail or Northern Bobwhite *Colinus virginianus* to be Georgia's state game bird.

The Joint Resolution said 'countless Georgians' had always considered the Brown Thrasher to be the state bird, so it was only 'fitting and proper' that this species 'be given the recognition it is due'.

The Resolution added that the Bobwhite Quail had 'charmed Georgians and accompanied them in their work and play since the state was merely a territory occupied by British colonists in 1733'. Despite the fact that 'thousands of Georgia sportsmen annually trek to the fields to bag their limit of Bobwhite Quail, this marvellous bird can withstand a loss of two-thirds of its population with no reduction in the spring breeding population, thus providing Georgia huntsmen with continued exciting sport.'

Hawaiian Goose

HAWAII

Hawaiian Goose *Branta sandvicensis*

Called the Nene (pronounced 'nay-nay') by Hawaiians, this species once teetered on the brink of extinction. Thanks to conservation measures, numbers have risen from perhaps just 30 birds in the mid-1900s to more than 2,000 individuals in 2011, according to BirdLife International. The Hawaiian Goose is currently designated as Vulnerable.

Writing in the June 1956 edition of the journal of the Hawaiian Audubon Society (*Elepaio*), Margaret Titcomb said if the Hawaiian Goose were to be saved from extinction, 'it will be a triumph of present-day conservation effort and technique. As long ago as 1903, and perhaps earlier, R. C. L. Perkins expressed fear and regret that it would disappear.'

Titcomb said the decline of this species had been attributed to livestock and other animals, changes in habitat and hunting. She referred to a visiting writer, Boddam-Whetham, who had described a strawberry-fed goose as having been and baked in a hole in the ground.

Western Meadowlark

Northern Cardinal

American Goldfinch

The Nene was declared to be the bird emblem of Hawaii in 1957.

IDAHO

Mountain Bluebird *Sialia currucoides*
Idaho has adopted two birds: the Mountain Bluebird (designated as the official state bird in 1931) and the Peregrine Falcon *Falco peregrinus* (official state raptor or bird of prey).

ILLINOIS

Northern Cardinal *Cardinalis cardinalis*
Schoolchildren selected the Cardinal to be Illinois's state bird in 1928 from a list of five conspicuous species. It was formally adopted as the official state bird the following year after a poll in which it attracted 39,226 votes. Illinois, in fact, was the first of seven US states to choose the Cardinal as its avian emblem, the others being Indiana, Kentucky, North Carolina, Ohio, Virginia and West Virginia.

The Cardinal was named by early settlers after Catholic Cardinals who wear striking red robes.

INDIANA

Northern Cardinal *Cardinalis cardinalis*
This familiar, colourful species (the male is bright red) was adopted as Indiana's official bird by the state's General Assembly in 1933 – possibly on account of the fact that it occurs in a wide range of habitats, including gardens.

Sports teams at Ball State University in Muncie, Indiana, are known as the Cardinals.

IOWA

American Goldfinch *Carduelis tristis*
Also known as the 'wild canary' because of the male's striking lemon plumage, the Goldfinch was chosen as Iowa's official bird in 1933 because it is commonly found in the state and often stays in the winter.

Its image is used on bookmarks and other items about Iowa's official symbols.

KANSAS

Western Meadowlark *Sturnella neglecta*
This species was designated Kansas' official state bird in 1937.

KENTUCKY

Northern Cardinal *Cardinalis cardinalis*
The Cardinal was chosen as Kentucky's official bird on 26 February 1926 simply because it is a native of the state.

Several elementary and high schools and one major university in Kentucky have adopted the Cardinal as their mascot.

LOUISIANA

Brown Pelican *Pelecanus occidentalis*
Although officially designated the state bird in 1966, the Brown Pelican's links with Louisiana date back centuries, in fact to the early European settlers who were impressed with this species' caring attitude towards its offspring.

Louisiana is nicknamed The Pelican State and the image of this bird appears on the state flag, state seal and state painting and is one of three Louisiana symbols featured on the USA's bicentennial quarter. Pesticides led to the complete disappearance of Brown Pelicans in Louisiana by 1966. Today, happily, there are around 40,000 of these iconic birds along the state's coastline.

MAINE

Black-capped Chickadee *Poecile atricapillus* A familiar garden bird, the Chickadee was adopted by the Maine Legislature as its state bird on 6 April 1927. The official citation mentions 'the chickadee' rather than any particular species, but it is assumed that the bird being referred to is the Black-capped

Brown Pelican

Baltimore Oriole

Black-capped Chickadee

Western Meadowlark

Chickadee, which can be seen and heard in Maine all year round.

Since 1999, Maine's general issue licence plate for cars has featured the image of a Chickadee on a tree branch. This bird is also used as the logo for the official Maine Birding Trail.

MARYLAND

Baltimore Oriole *Icterus galbula* This species was chosen as Maryland's avian icon in 1947 mainly because the male bird's black and golden-orange plumage resembles the colours in the state flag and Calvert family crest. (George Calvert, 1st Baron Baltimore, applied to Charles I for a royal charter to establish the Province of Maryland. The charter was actually granted to his son, Cecilius, on 20 June 1632, following his father' s death

Eastern Bluebird

earlier that year.)

In 1698, 'Baltemore Birds' were among 'Beasts of Curiosity' sent from Maryland to grace royal gardens.

Hoagy Carmichael composed the song *Baltimore Oriole* in the late 1930s (lyrics were written by Paul Francis Webster).

The Baltimore Oriole's image is most closely associated today with the Baltimore Orioles baseball team. A artoon image of the bird is used to express fan loyalty.

MASSACHUSETTS

Black-capped Chickadee *Poecile atricapillus* A common back-garden bird, the Chickadee was formally adopted as the state bird of Massachusetts in 1941.

The Wild Turkey was designated Massachusetts' official game bird in 1991.

MICHIGAN

American Robin *Turdus migratorious* Described in legislation as 'the best known and best loved' of all Michigan's birds, the American Robin was adopted as the official state bird in 1931 following an election by Michigan Audubon Society.

Efforts to change the state bird to Black-capped Chickadee or the rare and endangered Kirtland's Warbler have all been unsuccessful.

MINNESOTA

Common Loon *Gavia immer* Renowned for its eerie, far-carrying calls, the Common Loon was designated as Minnesota's state bird in 1961 after a successful bill was promoted by Senators Norman Walz and Loren Rutter and signed into law by Governor Elmer L. Andersen.

Other bird species suggested prior to 1961 included American Goldfinch, Mourning Dove, Pileated Woodpecker, Scarlet Tanager and Wood Duck.

MISSISSIPPI

Northern Mockingbird *Mimus polyglottos* This common species was officially designated Mississippi's state bird in 1944.

MISSOURI

Eastern Bluebird *Sialia sialis* Missouri adopted the Eastern Bluebird as its state bird on 30 March 1927. Common in the state from early spring until late November, the Bluebird is considered to be a symbol of happiness. Its image is depicted on a speciality car licence plate.

MONTANA

Western Meadowlark *Sturnella neglecta* Montana schoolchildren chose the Meadowlark in 1930 as the bird that best represented their state. The following year their choice was made official by the state legislature.

This species was first recorded for science on 22 June 1805 by explorer Meriwether Lewis in the Great Falls area of the Missouri River. He wrote; 'There is a kind of larke that much resembles the bird

Purple Finch

Common Loon

Greater Roadrunner

called the Oldfield Lark with a yellow brest and a black spot on the croop...The beak is somewhat longer and more curved and the note [song] differs considerably. However, in size, action and colours there is no perceptible difference, or at least none that strikes my eye.'

NEBRASKA

Western Meadowlark *Sturnella neglecta* On 25 October 1928, the Conservation Division of the Nebraska Federation of Women's Clubs, meeting a at convention in Kearney, resolved 'that a bird typical of the prairies and abundant in all parts of the state' be chosen as Nebraska's avian symbol. It was also decided that convention votes be added to the votes of Nebraskan schoolchildren and interested societies.

The five species which received the largest number of votes were Western Meadowlark, American Robin, Bobwhite, Brown Thrasher and House Wren. The Western Meadowlark was duly declared the vote of the people and a bill formally making the Meadowlark the state bird was signed by Governor Adam McMullen on 22 March 1929.

NEVADA

Mountain Bluebird *Sialia currucoides* This stunning high country species was chosen as Nevada's official state bird in 1967.

NEW HAMPSHIRE

Purple Finch *Carpodacus purpureus* The beautiful rose-coloured Purple Finch had to beat a chicken to become New Hampshire's state bird in 1957. Robert S. Monahan sponsored a Purple Finch bill, which he filed in the House of Representatives on 12 February of that year with solid backing. He later testified that his bill was supported by the Audubon Society of New Hampshire, the New Hampshire Federation of Garden Clubs and State Federation of Women's CLubs.

Veteran legislator Doris M. Spollett, however, wanted the New Hampshire Hen as state bird. Eight years earlier she had lost an initial bid to have this special breed of poultry chosen as New Hampshire's avian emblem.

Monahan urged speedy enactment for the Purple Finch proposal 'before some other state beats us to it'.

The Purple Finch attracted broad legislative support, given that it was backed by respected organisations. Governor Lane Dwinell signed the species into law as state bird on 25 April 1957.

Mountain Bluebird

NEW JERSEY

American Goldfinch *Carduelis tristis* This species was formally adopted as New Jersey's state bird in 1935.

NEW MEXICO

Greater Roadrunner *Geococcyx californianus* Believed by some Native American Indian tribes to be a protection against evil spirits, the Greater Roadrunner – also known as the Chaparral Bird – was officially designated New Mexico's state bird in 1949.

The Roadrunner's image is used on New Mexico's train system, the NM Rail Runner Express.

NEW YORK

Eastern Bluebird *Sialia sialis* This species became New York's official state bird in a law signed by Governor Nelson Rockefeller on 28 May 1970.

NORTH CAROLINA

Northern Cardinal *Cardinalis cardinalis* The campaign to choose an official state bird was initiated by North Carolina Bird Club and publicised through newspapers,

Western Meadowlark

Ruffed Grouse

Carolina Wren

birding/general wildlife clubs and schools. More than 23,000 votes were cast and 26 different bird species were put forward, including Red-winged Blackbird, Wild Turkey, Scarlet Tanager and Grey Catbird.

The Northern Cardinal won with 5,000 votes and became the official bird emblem of North Carolina on 8 March 1943.

Ten years earlier the state had a different avian icon – albeit very briefly. At the suggestion of the North Carolina General Federation of Women's Clubs, the General Assembly passed a resolution declaring the Carolina Chickadee to be the state's avian icon. A week later, however, the resolution was repealed because the Chickadee's nickname – Tomtit – was considered too undignified. Legislators were horrified at the prospect of North Carolina becoming known as the Tomtit State!

NORTH DAKOTA

Western Meadowlark *Sturnella neglecta*
The Meadowlark was chosen as North Dakota's official bird in 1947 by the Legislature because its melodic song is a very common sound on the open prairie which covers most of the state.

The bird's image is used on souvenirs, such as mugs, shirts and magnets – usually alongside the state flower, which is the Wild Prairie Rose.

OHIO

Northern Cardinal *Cardinalis cardinalis*
A common backyard bird, the Cardinal was chosen as Ohio's state bird in 1933.

OKLAHOMA

Scissor-tailed Flycatcher *Tyrannus forficatos* This striking, long-tailed bird – noted for its so-called sky dancing, which has been likened to an 'aerial ballet of incomparable grace' – became Oklahoma's state bird on 26 May 1951.

The campaign for the adoption of this species, which began with a group of schoolchildren, was supported by Oklahoma's Audubon Society, garden clubs, wildlife groups and Lou Allard, Chairman of the House Committee on Game and Fish.

The fact that the Scissor-tailed Flycatcher was of economic importance to farmers and ranchers (it eats lots of insects), its nesting range was centred on Oklahoma and no other state had chosen this species, all helped the Flycatcher to secure the designation of official state bird.

The Flycatcher is depicted in mid-flight on the back of the Oklahoma state quarter.

OREGON

Western Meadowlark *Sturnella neglecta*
The Meadowlark was chosen as Oregon's state bird by schoolchildren in 1927 in an Oregon Audubon Society poll.

Its image is used in various publications, including the *Oregon Blue Book* (almanac and fact book).

PENNSYLVANIA

Ruffed Grouse *Bonasa umbellus* The hardy Ruffed Grouse – one of 10 native, North American grouse species – was formally adopted as Pennsylvania's state bird in 1931.

RHODE ISLAND

Rhode Island Red This hen was adopted as Rhode Island's official bird on 3 May 1954 because this breed of poultry was developed in the state and in neighbouring Massachusetts. The Rhode Island Red is bred for its meat and eggs and ability to live in harsh conditions.

SOUTH CAROLINA

Carolina Wren *Thryothorus ludovicianus*
This species was made the state bird of South Carolina by Act No 693 of 1948 – the same piece of legislation which repealed an earlier Act designating the Northern Mockingbird as the state's avian symbol.

The Wild Turkey *Meleagris gallopavo* was declared the official wild game bird by Act No 508 of 1976.

SOUTH DAKOTA

Ring-necked Pheasant *Phasianus colchicus* Adopted as official state bird in 1943, the Ring-necked Pheasant is depicted on South Dakota's quarter coin.

California Gull

Hermit Thrush

Northern Cardinal

TENNESSEE

Northern Mockingbird *Mimus polyglottos* Following an election conducted by Tennessee Ornithological Society, the Northern Mockingbird was adopted as the state's official bird in 1933.

The Bobwhite Quail *Colinus virginianus* was adopted as Tennessee's official game bird in 1988.

TEXAS

Northern Mockingbird *Mimus polyglottos* Renowned for its vocal mimicry and huge repertoire of songs, the Northern Mockingbird was adopted as Texas' official state bird in 1927. It was noted in the legislation that this species was a 'singer of a distinctive type' and 'a fighter for the protection of its home.'

UTAH

California Gull *Larus californicus* This species was chosen as Utah's state bird in 1955 by an Act of Legislature. The Seagull Monument in Temple Square, Salt Lake City, honours this gull, which is noted for its aerial agility.

Northern Mockingbird

VERMONT

Hermit Thrush *Catharus guttatus* Vermont Federation of Women's Clubs adopted the Hermit Thrush as their state bird in 1927. Getting official recognition, however, took much longer. In fact, the Hermit Thrush had to wait until 1941 before it was officially declared to be Vermont's state bird.

One of the problems was that this species wasn't regarded as a true Vermonter because it migrates south in winter.

According to the Vermont Legislative Directory and State Manual, Biennial Session, 1993–1994, the Hermit Thrush was finally chosen because of its sweet call and it is found in all 14 of Vermont's counties.

VIRGINIA

Northern Cardinal *Cardinalis cardinalis* This species was chosen as Virginia's official state bird in 1950.

WASHINGTON

American Goldfinch *Carduelis tristis* In 1928, at the invitation of the Washington Legislature, schoolchildren selected the Western Meadowlark for adoption as state bird. Politicians, however, were concerned that this species had already been chosen by seven other states, including Oregon and Wyoming, so no action was taken at that time.

In 1931, the Washington Federation of Women's Clubs promoted a state-wide referendum, as a result of which the American Goldfinch (then known as the Willow Goldfinch) was identified as the most popular choice. It was not until 1951, however, that the Washington Legislature finally adopted the American Goldfinch as the state's official bird, although its status as such is 'not widely publicised or recognised,' according to the Washington Department of Fish and Wildlife.

WEST VIRGINIA

Northern Cardinal *Cardinalis cardinalis* The Northern Cardinal became West Virginia's state bird when House Resolution 12 was adopted on 7 March 1949.

WISCONSIN

American Robin *Turdus migratorious* Became the official state bird in 1949, having been voted for by Wisconsin's schoolchildren in 1926/27.

WYOMING

Western Meadowlark *Sturnella neglecta* This species was made Wyoming's state bird by the Nineteenth Legislature on 5 February 1927.

1982 State Bird Stamps

Each one of the 50 states comprising the USA has an avian and flower icon. All manner of diverse species are represented among the various state birds, from the Rhode Island Red chicken of Rhode Island and the Mountain Bluebird of Idaho and Nevada to the Greater Roadrunner of New Mexico and the Common Loon of Minnesota.

The commonest state birds are songbirds, the bright red Northern Cardinal adopted by no fewer than seven states and the Northern Mockingbird the official bird of five states being the most popular.

Most states chose their avian icon in the 1920s, '30s and '40s. The 'newest' state birds are New York's Eastern Bluebird and Georgia's Brown Thrasher, both of which declared as such in 1970.

ALABAMA
Northern Flicker

ALASKA
Willow Ptarmigan

ARIZONA
Cactus Wren

ARKANSAS
Northern Mockingbird

CALIFORNIA
California Quail

COLORADO
Lark Bunting

CONNECTICUT
American Robin

DELAWARE
Blue Hen Chicken

FLORIDA
Northern Mockingbird

GEORGIA
Brown Thrasher

HAWAII
Hawaiian Goose

IDAHO
Mountain Bluebird

ILLINOIS
Northern Cardinal

INDIANA
Northern Cardinal

IOWA
American Goldfinch

KANSAS
Western Meadowlark

KENTUCKY
Northern Cardinal

LOUISIANA
Brown Pelican

MAINE
Black-capped Chickadee

MARYLAND
Baltimore Oriole

MASSACHUSETTS
Black-capped Chickadee

MICHIGAN
American Robin

MINNESOTA
Common Loon

MISSISSIPPI
Northern Mockingbird

MISSOURI
Eastern Bluebird

MONTANA
Western Meadowlark

NEBRASKA
Western Meadowlark

NEVADA
Mountain Bluebird

NEW HAMPSHIRE
Purple Finch

NEW JERSEY
American Goldfinch

NEW MEXICO
Greater Roadrunner

NEW YORK
Eastern Bluebird

NORTH CAROLINA
Northern Cardinal

NORTH DAKOTA
Western Meadowlark

OHIO
Northern Cardinal

OKLAHOMA
Scissor-tailed Flycatcher

OREGON
Western Meadowlark

PENNSYLVANIA
Ruffed Grouse

RHODE ISLAND
Rhode Island Red

SOUTH CAROLINA
Carolina Wren

SOUTH DAKOTA
Ring-necked Pheasant

TENNESSEE
Northern Mockingbird

TEXAS
Northern Mockingbird

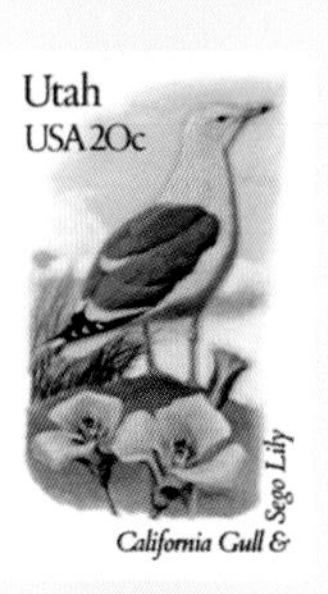

UTAH
California Gull

VERMONT
Hermit Thrush

VIRGINIA
Northern Cardinal

WASHINGTON
American Goldfinch

WEST VIRGINIA
Northern Cardinal

WISCONSIN
American Robin

WYOMING
Western Meadowlark

Venezuelan Troupial

Icterus icterus

LEAST CONCERN

IUCN: Precise numbers overall are unknown, but the Venezuelan Troupial is believed to be fairly common.

Size: Length 23–27 cm (9–10.6 in).

Description: Colourful New World blackbird, of which there are three subspecies. Nominate race is largely black and orange-yellow with white on its wings. Sexes are similar. All three subspecies – the nominate *I. i. icterus*, *I. i. metae* and *I. i. ridgwayi* – occur in Venezuela (and two also in Colombia).

Diet: Largely insects, fruits, nectar and seeds.

Reproduction: Generally three eggs per clutch are laid. Troupials are pirates in that they take over old nests made by other birds or drive the occupants away from active nests.

Range: Very large. Native to Aruba, Colombia, Netherlands Antilles, Puerto Rico, Venezuela and Virgin Islands.

Habitat: Nominate race is found mainly in pastures and woodland savanna.

The Venezuelan Troupial was formally adopted as Venezuela's national bird on 23 May 1958 following a two-month-long consultation process with various bodies and personalities conducted by the Venezuelan Society of Natural Sciences (SVCN) the previous year.

According to Luis Alberto Matheus of Audubon de Venezeula, the name *Troupial* is often used to describe something that is authentically Venezuelan. He also says the bird is frequently represented on public signs and murals along the major streets in countryside towns and schools. In the *llanos* in particular – the flat grassland covering much of the territory north of the Orinoco River and east of the Andes – the *Trupial* (a nickname for the bird) is common and often seen in backyards.

WHERE TO SEE The species can be seen from acacia and cacti woodlands to pastures, savannah grasslands and old fruit plantations. In coastal Venezuela, this species inhabits xerophytic woodland of acacia and aborescent cacti.

The Venezuelan Troupial has a loud clear song which is a series of two-part phrases: 'tree-trur' or 'cheer-tu'.

Above *In coastal Venezuela, the Troupial is found in acacia and cacti woodland.*

Right: *Perched atop a cactus, a dapper Venezuelan Troupial keeps a watchful eye for predators.*

Cultural presence

The Venezuelan Troupial was voted national bird in 1958. The four shortlisted species considered worthy of being crowned Venezuela's 'national emblematic bird', according to a 2009 paper by Eduardo Lopez, were the Venezuelan Troupial, the Andean Cock-of-the-rock *Rupicola peruviana*, the Guacharo, or Oilbird, *Steatornis caripensis* and 'the vulture'. The vulture, 'as always, vilified and discriminated unfairly', was the first to be eliminated in the final, followed by the Guacharo, which was 'considered scary by some'. The final choice, according to Lopez, was 'more fierce than expected', with arguments both for and against the two remaining species.

Critics of the Venezuelan Troupial said that this species also 'lived in captivity' and was 'adapted to the cage', although it was pointed out by a supporter that the horse had also been tamed by man and yet appeared on Venezuela's shield. In the end, the Venezuelan Troupial beat the Andean Cock-of-the-rock by 27 votes to 22. It was said to have melodious qualities, a pretty plumage and an aggressive flight.

Luis Alberto Matheus remarks that a song by Simon Diaz, a contemporary musician who has dedicated his life to saving dwindling music genres, incorporates a whistle that is a good imitation of the Venezuelan Troupial's song.

Regarded by at least some people as a herald of the dawn, the Venezuelan Troupial is mentioned in a poem by Hermes Delgado, who refers to having been awakened one morning by the 'sweet singing' of both a troupial and a goldfinch.

The Venezuelan Troupial appeared on a 5-cent stamp in a set of Venezuelan bird stamps in 1961. Its name has been widely used in brands for juice and fruit products and by tour operators, hotels, small inns, ranches and guest houses.

The Venezuelan Troupial was one of six species featured in a set of bird stamps in 1961.

Arabian Golden-winged Grosbeak

Rhynchostruthus percivali

NEAR THREATENED

IUCN: The total population is estimated to be around 9,000 individuals. It is thought that there are roughly 2,000 pairs in Yemen.

Size: Length 15 cm (5.9 in).

Description: Chunky finch with a large head and bill. Brown/grey-brown upperparts, a grey/black bill, white ear coverts and grey-buff underparts. Bright yellow secondary feathers. Sexes are broadly similar, although females are duller than males.

Diet: Mainly *Euphorbia* seeds, buds and fruits.

Reproduction: Very little known.

Range: Native to Oman, Saudi Arabia and Yemen.

Habitat: Frequents steep-sided valleys and coastal slopes in Oman, but its local movements are poorly known.

The Arabian Golden-winged Grosbeak was chosen as Yemen's national bird in 2008 by the Environment Minister, His Excellency Abdul Rahman Al-Eryani, following a lengthy consultation process.

WHERE TO SEE Places where this species has been recorded include the high mountains of Ibb, the Ta'izz wadis and the Mahra wooded escarpment, all of which are IBAs. BirdLife International says that this species is 'generally scarce and difficult to locate' even in those places where it is known to be present, 'with the exception perhaps of Yemen'. The Arabian Golden-winged Grosbeak species favours high altitude, scrub-carpeted rocky terrain.

This heavy-billed finch, which lives on seeds, buds and fruits, is found in Saudi Arabia, Oman and Yemen.

Cultural presence

Left: *Yemen has also adopted the Dragon Blood* Dracaena cinnabari *tree as its national tree, but this is only found on Socotra (which belongs to Yemen).*

Above: *It has adopted the Aloe* Aloe perryi *as its national plant.*

At the time of writing, no information was a available as to how the Arabian Golden-winged Grosbeak's image is being used in Yemeni society and how it will be used in official and other circles in the years ahead. This is largely because it only became Yemen's national bird in 2008.

At the same time, the country also named its national mammal (Arabian Leopard *Panthera pardus nimr*), tree (Dragon Blood Tree *Dracaena cinnabari*) and plant (*Aloe perryi*). In a statement to the Yemeni cabinet and media, the Environment Minister commented: 'I am proud we have chosen these animals and plants that are so important for Yemen's biodiversity and culture. They will help us promote wildlife education and conservation actions.'

Since 2004 BirdLife International has recognised the Arabian Golden-winged Grosbeak, Somali Golden-winged Grosbeak *R. louisae* and Socotra Golden-winged Grosbeak *R. socotranus* as three separate species. Before this, these species were lumped together under one common name: the Golden-winged Grosbeak.

The Arabian Golden-winged Grosbeak, appointed avian emblem in 2008. The coat of arms of Yemen is dominated by a 'Golden Eagle', depicted holding a scroll in its talons.

As its national mammal, Yemen has adopted the Arabian Leopard Panthera pardus nims.

African Fish-eagle

Haliaeetus vocifer

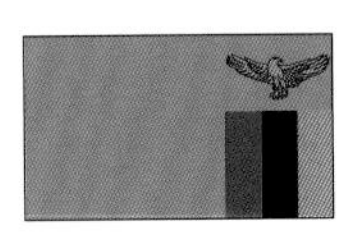

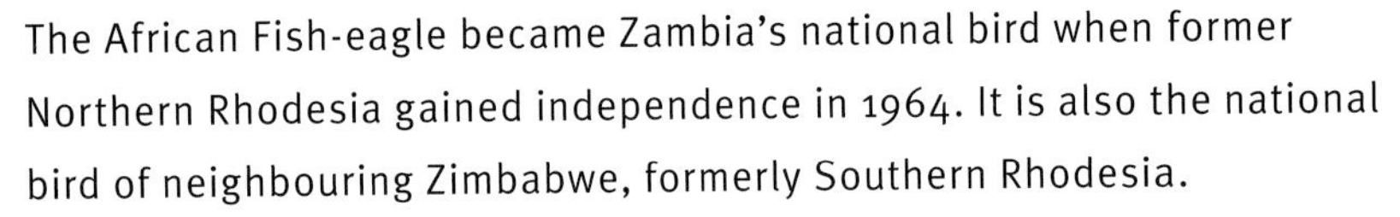

The African Fish-eagle became Zambia's national bird when former Northern Rhodesia gained independence in 1964. It is also the national bird of neighbouring Zimbabwe, formerly Southern Rhodesia.

The African Fish-eagle's image is widely used in Zambia, both in official circles and in commerce. The bird is shown in flight on the country's flag and coat of arms. It is also depicted – carrying a large fish – on the pre-independence flags of Northern Rhodesia and of the colonial governor of Northern Rhodesia.

WHERE TO SEE Found by major rivers and lakes throughout Zambia, including the Zambezi and Kafue Rivers, and Lakes Kariba and Tanganyika.

A good location to see an African Fish-eagle is the Chembe Bird Sanctuary on the Copperbelt, near Kitwe. The sanctuary is an area of wetland, grassland and woodland which surrounds a small lake. It is famous for its abundant birdlife, including the African Fish-eagle.

The area was formerly used as a reservoir to service local mining activity. It became a bird sanctuary in 1973 and has had official protection status as a national park for over 30 years.

LEAST CONCERN

IUCN: Very large population of 300,000 mature individuals, which appears to be stable.

Size: 63–75 cm (24.8–29.5 in). Medium-sized fishing eagle.

Description: Adults are a striking combination of white, chestnut-brown and black. Long, broad wings, a short, rounded tail and a large head with a prominent bill.

Diet: Mainly live fish, which it swoops down on and grabs from the water in its talons. Also takes nestlings and even some adult birds. Piratical, harrying other birds such as storks, herons and kingfishers into dropping their prey.

Reproduction: Two eggs are usually laid in a large stick nest in a tree near water.

Range: Found over much of sub-Saharan Africa to the most southerly parts of South Africa. Locally common to uncommon or rare.

Habitat: Wide-ranging. Lakes, rivers, floodplains and stocked dams inland. Estuaries, creeks and mangrove lagoons in coastal areas.

Two young African Fish-eagles on the look-out for thier next meal.

A Marabou Stork and an African Fish-eagle square up to one another over scraps of food.

Birds have been featured on many Zambian stamps over the years. This image of an African Fish-eagle appeared in 1975 on one of three bird definitives.

Cultural presence

A flying African Fish-eagle is shown on Zambia's flag and coat of arms.

Given that the African Fish-eagle is one of Africa's most familiar and photographed birds, it is unsurprising that various places and enterprises have been named after it. Among them is the Chembe Bird Sanctuary in Zambia's Copperbelt, where the species is often seen, *Chembe* being the local Lamba name for the bird. The African Fish-eagle has also lent its name to a brandy and to at least one safari business.

The species is a popular subject for artists and sculptors. In one of his paintings Canadian wildlife artist Robert Bateman shows the bird calling from the top of a tree. He says he was trying to convey 'the feeling of the powerful contortions of the African Fish-eagle's call through the visual rhythms of the gnarled trees'. Another artist, Guy Coheleach, depicts the African Fish-eagle against the backdrop of Victoria Falls.

The African Fish-eagle has appeared on various kwacha (Zambian currency) banknotes.

ZIMBABWE

African Fish-eagle

Haliaeetus vocifer

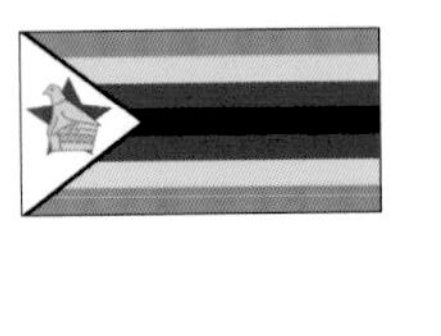

LEAST CONCERN

IUCN: Very large population of 300,000 mature individuals, which appears to be stable.

Size: 63–75 cm (24.8–29.5 in). Medium-sized fishing eagle.

Description: Adults are a striking combination of white, chestnut-brown and black. Long, broad wings, a short, rounded tail and a large head with a prominent bill.

Diet: Mainly live fish, which it swoops down on and grabs from the water in its talons. Also takes nestlings and even some adult birds. Piratical, harrying other birds such as storks, herons and kingfishers into dropping their prey.

Reproduction: Two eggs are usually laid in a large stick nest in a tree near water.

Range: Found over much of sub-Saharan Africa to the most southerly parts of South Africa. Locally common to uncommon or rare.

Habitat: Wide-ranging. Lakes, rivers, floodplains and stocked dams inland. Estuaries, creeks and mangrove lagoons in coastal areas.

The African Fish-eagle is the official national bird of Zimbabwe. It is an important national emblem, its image appearing frequently. However, according to some sources, the stylised Zimbabwe Bird is probably the African Fish-eagle, while others say it may be the Bateleur *Terathopius ecuadatus* – another eagle species. Whichever species the Zimbabwe Bird is meant to represent, the one clearly illustrated on the country's stamps is the African Fish-eagle.

The Zimbabwe Bird was also depicted on the flag of pre-Zimbabwe Southern Rhodesia.

WHERE TO SEE The African Fish-eagle is a largely sedentary, widely distributed bird and is therefore likely to be seen in a variety of aquatic habitats all year round.

The Bateleur, which is found across much of sub-Saharan Africa, including Zimbabwe, is declining as a result of habitat loss, poisoning and pollution, and is now classified as Near Threatened.

The African Fish-eagle's size and eye-catching plumage make it easy to see from a distance – especially in a skeletal tree.

Left: *A pair of African Fish-eagles. Although the sexes are similar, females have more white on their head and breast.*

Above: *A successful kill. An African Fish-eagle captured on camera at the moment it is about to pull away from the water carrying a fish in its talon.*

Cultural presence

The origins of the iconic Zimbabwe Bird date back to the ancient Shona city of Great Zimbabwe, in which 10,000 to 20,000 people are believed to have lived from the 11th to the 15th centuries. The impressive stone ruins of Great Zimbabwe – a World Heritage Site – can be seen near Masvingo in south-eastern Zimbabwe.

Eight birds carved from soapstone were found on columns within Great Zimbabwe when the site was excavated by archaeologists in the late 19th century. It has been suggested by academics that these Zimbabwe Birds symbolised royal authority. Shona people have long been noted for their bird and other stone art – a tradition continued by craftspeople in modern-day Zimbabwe. The Bateleur or *Chapungu* is considered by Shona people to be a bird of good omen and a messenger of the gods.

The Zimbabwe Bird is the logo of Air Zimbabwe founded in 1967 (formerly Air Rhodesia Corporation).

The African Fish-eagle was one of six raptors (birds of prey) featured on a set of Zimbabwe stamps in 1984.

The flag of Zimbabwe features the stylised Zimbabwe Bird based on the African Fish-eagle, or some claim the Bateleur.

Emblem miscellany

ALBANIA
Coat of arms

ARMENIA
Coat of arms

CANADA
Red-tailed Hawk

ALBANIA

A double-headed black eagle can be seen on Albania's flag.

ARMENIA

The coat of arms of Armenia incorporates an eagle.

BRITISH ANTARCTIC TERRITORY

An Emperor Penguin *Aptenodytes forsteri* is depicted on the coat of arms of the British Antarctic Territory.

CANADA

Although Canada does not have a national bird at present, concerted efforts are being made to have one adopted. Tens of thousands of people signed a petition organised by the Canadian Raptor Conservancy. A variety of birds was suggested by petitioners – species 'they feel most adequately' represent their country, according to Director James Cowan.

'Our hope is to take our petition to parliament and see if we can get a discussion or even a panel drawn up to actually choose a bird.'

Of those who signed the petition, around 70 per cent favoured the Red-tailed Hawk *Buteo jamaicensis*. The next most popular birds were Canada Goose *Branta canadensis*, the Common Loon or Great Northern Diver *Gavia immer* and the Grey Jay *Perisoreus canadensis*. Cowan cites the following reasons why it is thought that the Red-tailed Hawk would 'make a super national symbol':

1. It is found in every Canadian province and parts of every territory.
2. It is red – one of Canada's national colours.
3. 'We feel it is a ruggedly good-looking species – just like our country.'
4. It is an 'exclusively North American species, much like the Bald Eagle'.
5. It is a species of 'cultural significance' to Canada's First Nation peoples.
6. It is a 'highly visible bird that sits out in the open for all to witness – even in the midst of a city'.
7. It 'loves to sit in our national tree – the maple'.

The Canada Goose, adds Cowan, 'is a very polarising bird due to its grass eating – and eliminating – habits, and it also gets eaten a lot by the Americans' national symbol!' The Common Loon is already the provincial bird of Ontario – Canada's most populous province. The Grey Jay 'is a neat little bird but is very small and not often seen by the average Canadian as it is found further north in less populated areas'.

CANADIAN PROVINCES

Each province and territory has its own official bird emblem. They are: Alberta – Great Horned Owl *Bubo virginianus*; British Columbia – Steller's Jay *Cyanocitta stelleri*; Manitoba – Great Grey Owl *Strix nebulosa*; New Brunswick – Black-capped Chickadee *Parus atricapillus*; Newfoundland – Atlantic Puffin *Fratercula arctica*; Northwest Territories – Gyr Falcon *Falco rusticolus*; Nova Scotia – Osprey *Pandion haliaetus*; Nunavut – Rock Ptarmigan *Lagopus muta*; Ontario – Common Loon *Gavia immer*; Prince Edward Island – Blue Jay

CANADA
Canada Goose

CANADA
Common Loon

CANADA
Grey Jay

BRITISH COLOMBIA
Steller's Jay

NEW BRUNSWICK
Black-capped Chickadee

NUNAVUT
Rock Ptarmigan

MANITOBA
Great Grey Owl

NEWFOUNDLAND
Atlantic Puffin

PRINCE EDWARD ISLAND
Blue Jay

NOVIA SCOTIA
Osprey

NORTHWEST TERRITORIES
Gyr Falcon

QUEBEC
Snowy Owl

Cyanocitta cristata; Quebec – Snowy Owl *Bubo scandiaca*; Saskatchewan – Sharp-tailed Grouse *Tympanuchus phasianellus*; Yukon – Common Raven *Corvus corax*.

CHRISTMAS ISLAND

A golden Bosun Bird *Phaethon lepturus fulvus* – a distinct race of the White-tailed Tropicbird – is depicted on the flag of the Australian territory of Christmas Island.

CYPRUS

A dove carrying an olive branch in its bill is featured on Cyprus's coat of arms.

CZECH REPUBLIC

A red-and-silver chequered eagle is used on the coat of arms of Moravia in the Czech Republic, while a black eagle is shown on the coat of arms of Silesia.

EGYPT

The Eagle of Saladin holding a scroll is featured on the Egyptian coat of arms.

FIJI

If there is one bird that really ought to be a national bird, it is the Fiji Petrel *Pseudobulweria macgillivrayi*. Why? Because so much work has been done in Fiji to raise the profile of this Critically Endangered species in order to help save it.

Known locally as Kacau ni Gau, the Fiji Petrel has only ever been found on Gau – Fiji's fifth biggest island. According to the National Trust of Fiji's website (www.nationaltrust.org.fj), fewer than 10 sightings have been made since the first in 1855. BirdLife International estimates that the number of mature individuals could be fewer than 50. It is assumed that the tiny population is declining due to possible predation by cats, habitat destruction, bush fires, agricultural development and the activities of wild pigs.

A year-long community awareness campaign conducted in 2003 by the National Trust of Fiji in association with RARE and the National Parks and Wildlife

EGYPT
The Eagle of Saladin

FIJI
Fiji Petrel

CYPRUS
Dove

GERMANY
Stylised eagle

GHANA
Coat of arms

GUINEA
Stylised dove

IRELAND
Wren

Department of New South Wales included school and village visits on Gau Island, the production of brochures and posters, and a sermon and fact sheet. Businesses helped to fund outreach activities for schools. A group, Voqa ni Delaidokidoki, came up with a song about the life of the Fiji Petrel, the lyrics of which appealed for the bird to be protected. Gau chiefs agreed to set aside part of their highland forest as a reserve for conserving the species.

At 30 cm (11.8 in) in length, this small bird is dark chocolate-brown to black. It is thought to visit Gau only during the breeding season, spending the rest of the year far out at sea. The bird is featured on a Fijian banknote.

GERMANY
A stylised eagle is used on the country's coat of arms.

GHANA
The country's coat of arms is held by two 'golden eagles'.

GUINEA
A dove with a golden olive branch in its bill can be seen on the coat of arms of Guinea.

IRAQ
A gold and black eagle is featured on the Iraqi coat of arms.

IRELAND
Does not have a national bird, but BirdWatch Ireland says the Wren *Troglodytes troglodytes* has been suggested as a suitable candidate.

ISLE OF MAN
Supporting the coat of arms of the Isle of Man are a Peregrine Falcon *Falco peregrinus* and a Common Raven *Corvus corax*.

MALI
A golden vulture flying above a mosque is the centrepiece of Mali's coat of arms.

ISLE OF MAN
Coat of arms

ISLE OF MAN
Peregrine

MOLDOVA
Black Stork

POLAND
Stylised eagle

MOLDOVA

Although Moldova does not appear to have a designated national bird, both Black and White Storks *Ciconia nigra* and *C. ciconia* appear on bottles of brandy. A Black Stork was also depicted on a 2003 10-lei coin. The country's flag incorporates the Moldovan coat of arms featuring what is described as a 'golden eagle' holding an Orthodox Christian cross in its bill, and an olive branch and what seems to be a mace in its talons.

POLAND

Eagles have long been used in official circles in Poland. A stylised white eagle is depicted on the country's coat of arms, and there is an Order of the White Eagle for valour. Eagles also appear on military insignia.

According to Antoni Marczewski, a Polish politician, eagles are widely respected in Poland and 'people rather like them'.

ROMANIA

What is described as a golden *Aquila* dominates Romania's coat of arms. The eagle is shown holding a cross in its bill, and a mace and sword in its talons.

RUSSIAN FEDERATION

A double-headed eagle is featured on the coat of arms of the Russian Federation.

SABA

An Audubon's Shearwater *Puffinus lherminieri* is shown on the coat of arms of the Caribbean island of Saba.

ROMANIA
Stylised golden eagle

RUSSIAN FEDERATION
Stylised double-headed eagle

SABA
Audubon's Shearwater

SOUTH GEORGIA
Macaroni Penguin

SAINT BARTHÉLEMY

Two pelican-like birds are depicted on the flag and coat of arms of Saint Barthélemy – an overseas French collective in the Caribbean.

S O TOMÉ and PRINCIPÉ

A parrot and a falcon can be seen on the flag of São Tomé and Principé.

SOUTH GEORGIA

The Macaroni Penguin *Eudyptes chrysolophus* is depicted on the coat of arms of South Georgia and the South Sandwich Islands.

SOUTH SUDAN

The African Fish-eagle *Haliaeetus vocifer* forms the centrepiece of the coat of arms of South Sudan.

SYRIA

What is described as the Hawk of Qureish is depicted on Syria's coat of arms.

TURKEY

Although there are Internet references to the Redwing *Turdus iliacus* being Turkey's national bird, Sureyya Isfendiyaroglu, Conservation Director of Doga Dernegi, BirdLife International's partner designate for Turkey, says 'there is no such thing' as a national bird of Turkey. 'However, most Turkish and Anatolian folk had close associations with eagle species, especially the Golden Eagle *Aquila chrysaetos*.'

Falconry with Sparrowhawks *Accipiter nisus* is practised in some parts of Turkey, 'where this species is a commonly recognised bird. Some other bird species often referred to in classic and modern Turkish literature are cranes, chukar, flamingo, mallard, gulls, shearwaters, shelduck, nightingale, goldfinch and quail, but it's hard to call any of them national birds.' The Redwing is a winter migrant in Turkey and is 'barely recognised by people other than birdwatchers'.

UNITED ARAB EMIRATES

The UAE's coat of arms incorporates a golden falcon.

URUGUAY

The Southern Lapwing *Vanellus chilensis*, a familiar bird in Uruguay, is the mascot of the nation's Los Teros national rugby union team.

UZBEKISTAN

The holy Khumo bird – a symbol of love and freedom in Uzbek mythology – lies at the heart of the national emblem of Uzbekistan. The large, raptor-like bird is shown with its wings open.

SOUTH SUDAN
African Fish-eagle

TURKEY
Redwing

URURGUAY
Southern Lapwing

References

ANGOLA

BirdLife International species fact sheet on the Red-crested Turaco. www.birdlife.org/datazone

Handbook of the Birds of the World, volume 4.

Private correspondence with Michael Mills of Birds Angola.

ANGUILLA

BirdLife International species fact sheet on the Zenaida Dove.

www.birdlife.org/datazone

Handbook of the Birds of the World, volume 4.

ANTIGUA and BARBUDA

BirdLife International species fact sheet on the Magnificent Frigatebird. www.birdlife.org/datazone

Handbook of the Birds of the World, volume 1.

ARGENTINA

BirdLife International species fact sheet on the Rufous Hornero.

www.birdlife.org/datazone

Handbook of the Birds of the World, volume 8.

Wikipedia.

AUSTRALIA

BirdLife International species fact sheet on the Emu. www.birdlife.org/datazone

Handbook of the Birds of the World, volume 1.

Private correspondence with Andy Greenslade, Curator, Aboriginal and Torres Strait Islander Programs, National Museum of Australia, Canberra. www.nma.gov.au

Australian Museum. www.australianmuseum.net.au

Aboriginal Art Treasures. www.aboriginalarttreasures.com

Wikipedia.

Australian Government, Department of Foreign Affairs and Trade (coat of arms). www.dfat.gov.au

Australian fauna. www.australianfauna.com

AUSTRIA

BirdLife International species fact sheet on the Barn Swallow.

www.birdlife.org/datazone

Handbook of the Birds of the World, volume 9.

Private correspondence with Jutta Molterer of the Federal Ministry of Agriculture, Forestry, Environment and Water Management.

BAHAMAS

BirdLife International species fact sheet on the Caribbean or West Indian Flamingo. www.birdlife.org/datazone

Handbook of the Birds of the World, volume 1.

The Bahamas National Trust. www.bnt.bs

BAHRAIN

BirdLife International species fact sheet on the White-eared or White-eared Bulbul. www.birdlife.org/datazone

Handbook of the Birds of the World, volume 10.

BANGLADESH

BirdLife International species fact sheet on the Oriental Magpie-robin. www.birdlife.org/datazone

Handbook of the Birds of the World, volume 10.

BELARUS

BirdLife International species fact sheet on the White Stork.

www.birdlife.org/datazone

Handbook of the Birds of the World, volume 1.

'Preservation of White Stork population in Belarusian Polessia', by Irina Samusenko (Minsk 2000). MAB Young Scientist research Project Final report.

Wikipedia.

BELGIUM

Private correspondence with Joris Gansemans of Natuurpunt. www.natuurpunt.be

BELIZE

BirdLife International species fact sheet on the Keel-billed Toucan. www.birdlife.org/datazone

Handbook of the Birds of the World, volume 7.

Chaa Creek Nature Reserve. www.chaacreek.com

BERMUDA

BirdLife International species fact sheet on the Bermuda Petrel or Cahow. www.birdlife.org/datazone

Handbook of the Birds of the World, volume 1.

'The Fabled Cahow', by Dr David Wingate, Chief Conservation Officer, Bermuda Zoological Society, December 2006. Published by Bermuda Zoological Society as an 'ecofile information document'.

Private correspondence with Andrew Dobson, President, Bermuda Audubon Society. www.audubon.bm

The Cahow Recovery Program, Government of Bermuda (Ministry of the Environment and Planning, Department of Conservation Services). www.conservation.bm

'Cahow chick first to be born on Nonsuch Island since 1620', Government of Bermuda (The Cabinet Office, Department of Communication and Information), 16 April 2009.

Private correspondence with Jeremy Madeiros and Andrew Dobson.

BHUTAN

Birdlife International species fact sheet on the Common Raven.

www.birdlife.org/datazone

Handbook of the Birds of the World, volume 14.

Destination Bhutan: National Bird. www.bhutantrip.weebly.com

BOLIVIA

BirdLife International species fact sheet on the Andean Condor.

www.birdlife.org/datazone

Handbook of the Birds of the World, volume 2.

Bolivia: National Order of the Condor of the Andes. www.medals.org.uk

Wildlife Conservation Society (WCS Bolivia). www.wcsbolivia.org

BOTSWANA

BirdLife International species fact sheets on the Lilac-breasted Roller and Kori Bustard. www.birdlife.org/datazone

Handbook of the Birds of the World, volumes 3 and 6.

Private correspondence with Dr Kabelo Senyatso, Director, Birdlife Botswana. www.birdlifebotswana.org.bw

BRAZIL

BirdLife International species fact sheet on the Rufous-bellied Thrush. www.birdlife.org/datazone

Handbook of the Birds of the World, volume 10.

National symbols, Brazil Embassy. www.brazilembassy.org

BRITISH VIRGIN ISLANDS

BirdLife International species fact sheet on the Zenaida Dove.

www.birdlife.org/datazone

Handbook of the Birds of the World, volume 4.

History and Culture, Government of the British Virgin Islands. www.bvi.gov.vg

CAMBODIA

BirdLife International species fact sheet on the Giant Ibis.

www.birdlife.org/datazone

Handbook of the Birds of the World, volume 1.

Tmatboey Ecotourism/Wildlife Friendly Ibis Rice/Bird Nest Protection in the Northern Plains of Cambodia. Wildlife Conservation Society, Cambodia. www.wcscambodia.org

CAYMAN ISLANDS

BirdLife International species fact sheet on the Grand Cayman Parrot and Cayman Brac Parrot.

www.birdlife.org/datazone

Handbook of the Birds of the World, volume 4.

National Trust for the Cayman Islands. www.nationaltrust.org.ky

Private correspondence with Paul Watler, Environmental Programmes Manager, National Trust for the Cayman Islands.

Cayman Islands National Symbols: Flora and Fauna – National Bird.

CHILE

BirdLife International species fact sheet on the Andean Condor.

www.birdlife.org/datazone

Handbook of the Birds of the World, volume 2.

thisisChile. www.thisischile

Wikipedia.

CHINA

BirdLife International species fact sheet on the Red-crowned Crane. www.birdlife.org/datazone

Handbook of the Birds of the World, volume 3.

International Crane Foundation. www.savingcranes.org

Zhalong National Nature Reserve. www.scwp.info/china/zhalong

COLOMBIA

BirdLife International species fact sheet on the Andean Condor.

www.birdlife.org/datazone

Handbook of the Birds of the World, volume 2.

'The Andean Condor: A Field Study', by Jerry McGhan (study conducted from 1968 to 1970 for a graduate degree).

San Diego Zoo. www.sandiegozoo.org

Wikipedia.

COSTA RICA

BirdLife International species fact sheet on the Clay-coloured Thrush. www.birdlife.org/datazone

Handbook of the Birds of the World, volume 10.

National Bird of Costa Rica. www.1000birds.com

Costaricatourism. www.CostaRicatourism.co.cr

CUBA

BirdLife International species fact sheet on the Cuban Trogon.

www.birdlife.org/datazone

Handbook of the Birds of the World, volume 6.

Cuba's National Bird. www.circuloguinero.org

DENMARK

Birdlife International species fact sheet on the Mute Swan.

www.birdlife.org/datazone

Handbook of the Birds of the World, volume 1.

Private correspondence with Vicky Knudsen of Dansk Ornitologisk Forening (BirdLife Denmark). www.dof.dk

National Bird of Denmark. www.fyidenmark.com

DOMINICA

BirdLife International species fact sheet on the Imperial Amazon.

www.birdlife.org/datazone

Handbook of the Birds of the World, volume 4.

Private correspondence with Dr Paul R. Reillo, Director, Rare Species Conservatory Foundation. www.rarespecies.org

Forestry, Wildlife and Parks Division, Ministry of Agriculture and the Environment, Dominica. www.avirtualdominica.com/forestry/

'Parrot Conservation on Dominica: Successes, Challenges and Techological Innovations', by Paul R. Reillo and Stephen Durand, *Journal of Caribbean Ornithology* 21: 52–58, 2008.

'Dominica's National Bird – Sisserou Parrot/Two Sisserou Chicks from One Nest'. Government of the Commonwealth of Dominica. www.dominica.gov.dm

'First Captive Breeding of the Imperial Parrot', by Paul R. Reillo, Stephen Durand and Minchinton Burton, *Zoo Biology* 30: 328–341 (2011).

World Parrot Trust. www.parrots.org

DOMINICAN REPUBLIC

BirdLife International species fact sheet on the Palmchat.

www.birdlife.org/datazone

Handbook of the Birds of the World, volume 10.

ECUADOR

BirdLife International species fact sheet on the Andean Condor.

www.birdlife.org/datazone

Handbook of the Birds of the World, volume 2.

EL SALVADOR

BirdLife International species fact sheet on the Turquoise-browed Motmot. www.birdlife.org/datazone

Handbook of the Birds of the World, volume 6.

ESTONIA

BirdLife International species fact sheet on the Barn Swallow.

www.birdlife.org/datazone

Handbook of the Birds of the World, volume 9.

National symbols of Estonia. www.estinst.ee

Wikipedia.

FAROE ISLANDS

BirdLife International species fact sheet on the Eurasian Oystercatcher. www.birdlife.org/datazone

Handbook of the Birds of the World, volume 3.

FINLAND

BirdLife International species fact sheet on the Whooper Swan.

www.birdlife.org/datazone

Handbook of the Birds of the World, volume 1.

Private correspondence with Jan Sodersved, Communications Officer, BirdLife Finland. www.birdlife.fi

Private correspondence with Jukka Haapala, Ringing Centre, Finnish Museum of Natural History. www.luomos.fi/english/zoology/ringing

'The Swan is the theme of the new gold coin'. Bank of Finland. www.suomenrahapaja.fi.eng

Wikipedia.

FRANCE

www.123independenceday.com

Wikipedia – Gallic Rooster,1998 FIFA World Cup, Le Coq Sportif and France's national rugby team

www.franceway.com - Republic symbols

www.wisegeek.com - Gallic Rooster

GIBRALTAR

BirdLife International species fact sheet on the Barbary Partridge. www.birdlife.org/datazone

Handbook of the Birds of the World, volume 2.

Private correspondence with Tracey Poggio, UK Marketing and Communications, Gibraltar Tourist Board (London Office). www.gibraltar.gov.uk

GRENADA

BirdLife International species fact sheet on the Grenada Dove.

www.birdlife.org/datazone

Handbook of the Birds of the World, volume 4.

Grenada Dove – National Bird of Grenada. www.gov.gd

Grenada Dove, Wikipedia.

Coat of Arms of Grenada. www.gov.gd

GUATEMALA

BirdLife International species fact sheet on the Resplendent Quetzal. www.birdlife.org/datazone

Handbook of the Birds of the World, volume 6.

Order of the Quetzal/Guatemalan Quetzal/Resplendent Quetzal, Wikipedia.

'Uncanny Acoustic Effects at Chichen Itza: Intentional Design?', by David Lubman, The Acoustics of Ancient Theatres Conference, Patras, 18–21 September 2011.

Private correspondence with David Lubman.

GUYANA

BirdLife International species fact sheet on the Hoatzin.

www.birdlife.org/datazone volume 3.

Private correspondence with Annarie Shiwram, Product Development Officer, Guyana Tourism Authority. www.guyana-tourism.com

Guana Birding. www.guyanabirding.com

Wikipedia.

HAITI

BirdLife International species fact sheet on the Hispaniolan Trogon. www.birdlife.org/datazone

Handbook of the Birds of the World, volume 6.

HONDURAS

BirdLife International species fact sheet on the Scarlet Macaw.

www.birdlife.org/datazone

Handbook of the Birds of the World, volume 4.

Macaw Mountain Bird Park and Nature Reserve. www.macawmountain.com

World Parrot Trust. www.parrots.org

Born Free USA. www.bornfreeusa.org

HUNGARY

BirdLife International species fact sheet on the Great Bustard.

www.birdlife.org/datazone

Handbook of the Birds of the World, volume 3.

'The Great Bustard in Hungary', by Gergely Karoly Kovacs, Hungarian Ornithological and Nature Conservation Society. www.mme.hu

Private correspondence with Gergely Karoly Kovacs.

ICELAND

BirdLife International species fact sheet on the Gyrfalcon.

www.birdlife.org/datazone

Handbook of the Birds of the World, volume 2.

Wikipedia.

The Peregrine Fund. www.peregrinefund.org

'Gyrfalcon Population and Reproduction in Relation to Rock Ptarmigan Numbers in Iceland', by Olafur K. Nielsen. Online paper.

Private correspondence with Vilhjalmur Sigurdsson, Head of Philately, Iceland Post.

Order of the Falcon, Wikipedia.

INDIA

BirdLife International species fact sheet on the Indian Peafowl.

www.birdlife.org/datazone

Handbook of the Birds of the World , volume 2.

Wikipedia.

INDONESIA

BirdLife International species fact sheet on the Javan Hawk-eagle.

www.birdlife.org/datazone

Handbook of the Birds of the World, volume 2.

Private correspondence with ProFauna. www.profauna.net

Javan Hawk-eagle, Global Raptor Information Network, The Peregrine Fund. www.globalraptors.org

'Declaration of the Javan Hawk-eagle as Indonesia's Rare Animal Impedes Conservation of the Species', by Vincent Nijman, Chris Shepherd and S. van Balen, *Oryx*, 2009.

ISRAEL

BirdLife International species fact sheet on the Eurasian Hoopoe.

www.birdlife.org/datazone

Handbook of the Birds of the World, volume 6.

Private correspondence with Professor Yossi Leshem of the George S. Wise Faculty of Life Sciences, Department of Zoology, Tel Aviv University, Tel Aviv.

JAMAICA

BirdLife International species fact sheet on the Red-billed Streamertail.

www.birdlife.org/datazone

Handbook of the Birds of the World, volume 5.

'Colourful Characters – Jamaica's Birds', by Dr Rebecca Tortello.

www.jamaica-gleaner.com

The Doctor Bird. www.1000birds.com

Doctor Bird. www.jamaicatravelandculture.com

JAPAN

BirdLife International species fact sheet on the Green Pheasant.

www.birdlife.org/datazone

Handbook of the Birds of the World, volume 2.

Green Pheasant: The National Bird of Japan. www.thelovelyplanet.net

Private correspondence with Seiji Hayama of the Wild Bird Society of Japan. www.wbsj.org

JORDAN

BirdLife International species fact sheet on the Pale or Sinai Rosefinch.

www.birdlife.org/datazone

Handbook of the Birds of the World, volume 15.

Private correspondence with Tareq Qaneer of the Royal Society for the Conservation of Nature (RSCN).

LATVIA

BirdLife International species fact sheet on the White Wagtail.

www.birdlife.org/datazone

Handbook of the Birds of the World, volume 9.

National symbols of Latvia. www.latvia.lv

Private correspondence with Viesturs Kerus, Chief Executive, Latvian Ornithological Society (LOB), and also Dr Maris Strazds. www.lob.lv

LIBERIA

BirdLife International species fact sheet on the Common Bulbul.

www.birdlife.org/datazone

Handbook of the Birds of the World, volume 10.

Private correspondence with Michael Garbo of the Society for the Conservation of Nature in Liberia (SCNL).

LITHUANIA

BirdLife International species fact sheet on the White Stork.

www.birdlife.org/datazone

Handbook of the Birds of the World, volume 1.

Private correspondence with the Lithuanian Ornithological Society (LOD). www.birdlife.lt

LUXEMBOURG

BirdLife International species fact sheet on the Goldcrest.

www.birdlife.org/datazone

Handbook of the Birds of the World, volume 11.

MALAWI

BirdLife International species fact sheet on the African Fish-eagle. www.birdlife.org/datazone

Handbook of the Birds of the World, volume 2.

Lake Malawi National Park, UNESCO World Heritage Centre. www.whc.unesco.org

MALTA

BirdLife International species fact sheet on the Blue Rock-thrush. www.birdlife.org/datazone

Handbook of the Birds of the World, volume 10.

LIFE+ Project on Bird Migration and Trapping, BirdLife Malta/RSPB/Media Today, November 2009.

MAURITIUS

Natural History Museum, Mauritius. www.mauritiusmuseums.com

Ile aux Aigrettes Nature Reserve. www.ile-aux-aigrettes.com

Extinct Birds, by Dr Julian P. Hume and Michael Walters, 2012, T & A D Poyser (Bloomsbury plc).

Private correspondence with Dr Julian P. Hume.

The Dodo Today. www.dodosite.com

MEXICO

BirdLife International species fact sheet on the Crested Caracara. www.birdlife.org/datazone

Handbook of the Birds of the World, volume 2.

MONGOLIA

http://ubpost.mongolnews.mn/?p=1863 - Saker Falcon, national bird of Mongolia, is it perishing by thousands?

www.birdlife.org/datazone
www.infomongolia.com – The National Bird of Glory – Falcon and Falcon Has Been Avowed as the National Pride Bird of Mongolia

Southern Connecticut State University – Counting falcons

www.southernct.edu

Handbook of the Birds of the World, volume 2

Developing A Sustainable Harvest of Saker Falcons (Falco cherrug) for Falconry in Mongolia by Andrew Dixon, Nyambayar Batbayar, Gankhuyag Purev-Ochir and Nick Fox. Global Raptor Information Network www.globalraptors.org

MONTSERRAT

BirdLife International species fact sheet on the Montserrat Oriole. www.birdlife.org/datazone

Handbook of the Birds of the World, volume 16.

Montserrat Oriole Species Action Plan, 2005–2009.

Montserrat Oriole, Durrell Wildlife Conservation Trust, March 2001.

www.durrell.org

'Demography and Breeding Ecology of the Critically Endangered Montserrat Oriole', by R. Allcorn, G. Hilton, C. Fenton, P. Atkinson, C. Bowden, C. Gray, M. Hulme, J. Madden, E. Mackley and S. Oppel, The Condor 114 (1):227–235, The Cooper Ornithological Society 2012.

Private correspondence with Steffen Oppel, Senior Conservation Scientist, RSPB.

MYANMAR

BirdLife International species fact sheet on the Grey Peacock-Pheasant. www.birdlife.org/datazone

Handbook of the Birds of the World, volume 2.

NAMIBIA

BirdLife International species fact sheet on the African Fish-eagle. www.birdlife.org/datazone

Handbook of the Birds of the World, volume 2.

NEPAL

BirdLife International species fact sheet on the Himalayan Monal or Danphe. www.birdlife.org/datazone

Handbook of the Birds of the World, volume 2.

Himalayan Nature. www.himalayannature.org

Nepalese Ornithological Union. www.birdsofnepal.org

Private correspondence with Dr Hem Baral, Chief Technical Adviser, Himalayan Nature, and Chairman, Nepalese Ornithological Union.

NEW CALEDONIA

BirdLife International species fact sheet on the Kagu.

www.birdlife.org/datazone

Handbook of the Birds of the World, volume 3.

NEW ZEALAND

BirdLife International species fact sheet on the Kiwi.

www.birdlife.org/datazone

Handbook of the Birds of the World, volume 1.

Kiwi Encounter, Rainbow Springs Kiwi Wildlife Park. www.rainbowsprings.co.nz

Otorohanga Kiwi House and Native Bird Park. www.kiwihouse.org.nz

BNZ Save the Kiwi Trust. www.savethekiwi.org.nz

Private correspondence with Toni Thompson and Claire Travers, The National Kiwi Trust at Kiwi Encounter, Rainbow Springs Kiwi Wildlife Park.

New Zealand's Flightless Birds Help Underprivileged Kids to Fly. www.stamps.nzpost.co.nz

Kiwi, Wikipedia.

Kiwi and People: Early History. www.teara.govt.nz

NICARAGUA

BirdLife International species fact sheet on the Turquoise-browed Motmot. www.birdlife.org/datazone

Handbook of the Birds of the World, volume 6.

NIGERIA

BirdLife International species fact sheet on the Black Crowned Crane. www.birdlife.org/datazone

Handbook of the Birds of the World, volume 3.

Nigerian Conservation Foundation. www.ncfnigeria.org

Private correspondence with Ruth Akagu, Senior Conservation Officer, Species and IBA Programme, Nigerian Conservation Foundation.

International Crane Foundation. www.savingcranes.org

Private correspondence with the International Crane Foundation.

NORWAY

BirdLife International species fact sheet on the White-throated Dipper. www.birdlife.org/datazone

Handbook of the Birds of the World, volume 10.

PAKISTAN

BirdLife International species fact sheet on the Chukar.

www.birdlife.org/datazone

Handbook of the Birds of the World, volume 2.

Chukar Partridge, Wikipedia.

PALAU

BirdLife International species fact sheet on the Palau Fruit-dove. www.birdlife.org/datazone

Handbook of the Birds of the World, volume 4.

Palau Conservation Society. www.palauconservation.org

Helm: London. Gibbs, D., Eustace, B., and Cox, J., (2001), *Pigeons and Doves: A Guide to the Pigeons and Doves of the World*.

PALESTINE

BirdLife International species fact sheet on the Palestine Sunbird. www.birdlife.org/datazone

Handbook of the Birds of the World, volume 13.

Palestinian Sunbird. Bethlehem Visitor Information Centre. www.vicbethlehem.wordpress.com

Private correspondence with Imad Atrash, Palestine Wildlife Society. www.wildlife-pal.org

PANAMA

BirdLife International species fact sheet on the Harpy Eagle. www.birdlife.org/datazone

Handbook of the Birds of the World, volume 2.

David Kitler, Eyes for the Wild Art. www.davidkitler.com

Private correspondence with Ly Kitler.

PAPUA NEW GUINEA

BirdLife International species fact sheet on the Raggiana or Count Raggi's Bird-of-Paradise. www.birdlife.org/datazone

Handbook of the Birds of the World, volume 14.

'Birds of Paradise: Revealing the World's Most Extraordinary Birds', by Tim Laman and Edwin Scholes, National Geographic Society and the Cornell Lab of Ornithology, 2012.

Flag of Papua New Guinea, Wikipedia.

Birds-of-Paradise, Wikipedia.

Baiyer River Wildlife Sanctuary. www.pngtourism.org.pg/Westernhighlands

PARAGUAY

BirdLife International species fact sheet on the Bare-throated Bellbird. www.birdlife.org/datazone

Handbook of the Birds of the World, volume 9.

FAUNA Paraguay, www.faunaparaguay.com

Private correspondence with Paul Smith of FAUNA Paraguay.

Bosque Mbaracayu Biosphere Reserve. www.unesco.org Search biosphere reserves directory.

PERU

BirdLife International species fact sheet on the Andean Cock-of-the-rock. www.birdlife.org/datazone

Handbook of the Birds of the World, volume 9.

PHILIPPINES

BirdLife International species fact sheet on the Philippine Eagle. www.birdlife.org/datazone

Handbook of the Birds of the World, volume 2.

Private correspondence with Tatit Quiblat, Manager for Development, Philippine Eagle Foundation. www.philippineeagle.org

PUERTO RICO

BirdLife International species fact sheet on the Puerto Rican Spindalis. www.birdlife.org/datazone

Handbook of the Birds of the World, volume 16.

Wikipedia.

Natural History Society of Puerto Rico (Sociedad de Historia Natural De Puerto Rico). www.shnpr.org

SAINT KITTS and NEVIS

BirdLife International species fact sheet on the Brown Pelican. www.birdlife.org/datazone

Handbook of the Birds of the World, volume 1.

Coat of Arms of St Kitts and Nevis, Wikipedia.

SAMOA

BirdLife International species fact sheet on the Tooth-billed Pigeon. www.birdlife.org/datazone

Handbook of the Birds of the World, volume 4.

'Saving the Biodiversity of Samoa: The Mao and Manumea Project'. www.samoanbirds.com

'Recovery Plan for the Manumea or Tooth-billed Pigeon', 2006–2016, Ministry of Natural Resources and Environment (MNRE), Government of Samoa. www.mnre.gov.ws

SCOTLAND

Birdlife International secies fact sheet on the Golden Eagle. www.birdlife.org/datazone

Handbook of the Birds of the World, volume 2.

RSPB: www.rspb.org.uk

Scottish Matural Heritage: snh.gov.uk
News Release 1 Nov 2013.

SEYCHELLES

BirdLife International species fact sheet on the Seychelles Black Parrot. www.birdlife.org/datazone

Handbook of the Birds of the World, volume 4.

Seychelles Islands Foundation. www.sif.sc

Nature Seychelles. www.natureseychelles.org

World Parrot Trust. www.parrots.org

Wikipedia.

SINGAPORE

BirdLife International species fact sheet on the Crimson Sunbird. www.birdlife.org/datazone

Handbook of the Birds of the World, volume 13.

Private correspondence with Alan Ow Yong, Nature Society (Singapore).

SOUTH AFRICA

BirdLife International species fact sheet on the Blue Crane. www.birdlife.org/datazone

Handbook of the Birds of the World, volume 3.

Kwande Private Game Reserve. www.kwande.com

Overberg Blue Crane Group. www.bluecrane.org.za

Endangered Wildlife Trust. www.ewt.org.za

International Crane Foundation. www.savingcranes.org

Private correspondence with Kerry Morrison, Manager, African Crane Conservation Programme, Endangered Wildlife Trust.

SRI LANKA

BirdLife International species fact sheet on the Sri Lanka Junglefowl. www.birdlife.org/datazone

Handbook of the Birds of the World, volume 2.

ST HELENA

BirdLife International species fact sheet on the St Helena Plover. www.birdlife.org/datazone

Handbook of the Birds of the World, volume 3.

St Helena National Trust Wirebird Conservation Programme. www.nationaltrust.org.sh

Private correspondence with Dr Chris Hillman, Wirebird Conservation Manager, St Helena National Trust.

ST LUCIA

BirdLife International species fact sheet on the St Lucia Amazon. www.birdlife.org/datazone

Handbook of the Birds of the World, volume 4.

World Parrot Trust. www.parrots.org

St Lucia Amazon/Parrot. Durrell Wildlife Conservation Trust. www.durrell.org

ST VINCENT

BirdLife International species fact sheet on the St Vincent Amazon. www.birdlife.org/datazone

Handbook of the Birds of the World, volume 4

St Vincent Amazon, Wikipedia.

World Parrot Trust. www.parrots.org

SUDAN

BirdLife International species fact sheet on the Secretarybird.

www.birdlife.org/datazone

Handbook of the Birds of the World, volume 2.

Emblem of Sudan, Wikipedia.

SWAZILAND

BirdLife International species fact sheet on the Purple-crested Turaco. www.birdlife.org/datazone

Handbook of the Birds of the World, volume 4.

Swaziland Tourism (UK). www.welcometoswaziland.com

Private correspondence with Kelly White of Swaziland Tourism (UK).

Private correspondence with Clive Humphreys of the International Turaco Society. www.turacos.co.uk

SWEDEN

BirdLife International species fact sheet on the Eurasian Blackbird. www.birdlife.org/datazone

Handbook of the Birds of the World, volume 10.

TAIWAN

BirdLife International species fact sheet on the Formosan Magpie. www.birdlife.org/datazone

Handbook of the Birds of the World, volume 14.

Taiwan International Birding Association. www.birdingtaiwan.com

THAILAND

BirdLife International species fact sheet on the Siamese Fireback. www.birdlife.org/datazone

Handbook of the Birds of the World, volume 2.

'Observations on the Siamese Fireback in Khao Yai National Park, Thailand', by Nichaya Praditsup, Amara Naksathit and Philip Round. *Forktail* 23 (2007): 125–128.

TRINIDAD and TOBAGO

BirdLife International species fact sheets on the Scarlet Ibis and Rufous-vented Chachalaca. www.birdlife.org/datazone

Handbook of the Birds of the World, volumes 1 and 2.

Caroni Bird Sanctuary. www.caronibirdsanctuary.com

Scarlet Ibis, Wikipedia.

Rufous-vented Chachalaca, Wikipedia.

UGANDA

BirdLife International species fact sheet on the Grey Crowned Crane. www.birdlife.org/datazone

Handbook of the Birds of the World, volume 3.

International Crane Foundation. www.savingcranes.org

Grey Crowned Crane, Durrell Wildlife Conservation Trust. www.durrell.org

Private correspondence with the International Crane Foundation.

UNITED KINGDOM

BirdLife International species fact sheet on the Eurasian Robin. www.birdlife.org/datazone

Handbook of the Birds of the World, volume 10

Wikipedia.

Christmas cards. www.whychristmas.com/customs/cards

Legendary Dartmoor. http://legendarydartmoor.co.uk/robin_moor.htm

Old Wives' Tales and Old Beliefs. www.bellaonline.com

US VIRGIN ISLANDS

BirdLife International species fact sheet on the Bananaquit. www.birdlife.org/datazone

Handbook of the Birds of the World, volume 16.

Virgin Islands National Park. www.nps.gov

St John message board of TripAdvisor. www.tripadvisor.co.uk

Virgin Islands. www.caribtourism.net/us-virgin-islands/flora-fauna

USA

BirdLife International species fact sheet on the Bald Eagle. www.birdlife.org/datazone

Handbook of the Birds of the World, volume 2.

Chilkat Bald Eagle Preserve. www.dnr.alaska.gov/parks/units/eagleprv.htm

American Bald Eagle Foundation. www.baldeagles.org

American Eagle Foundation. www.eagles.org

National Eagle and Wildlife Property Repository. www.fws.gov/le/Natives/EagleRepository.htm

Brackendale Winter Eagle Festival. www.brackendaleartgallery.com

State Symbols USA. www.statesymbols.org

VENEZUELA

BirdLife International species fact sheet on the Venezuelan Troupial. www.birdlife.org/datazone

Handbook of the Birds of the World, volume 16.

'Bird of the Month, May 2009'. Article by Eduardo Lopez. Published privately in Spanish.

Private correspondence with Luis Albert Matheus of Audubon de Venezuela.

YEMEN

BirdLife International species fact sheet on the Arabian Grosbeak. www.birdlife.org/datazone

Handbook of the Birds of the World, volume 15.

ZAMBIA

BirdLife International species fact sheet on the African Fish-eagle.www.birdlife.org/datazone

Handbook of the Birds of the World, volume 2.

Private correspondence with Zambia Ornithological Society. www.wattledcrane.com

Indigenous medicine (muthi) trade. www.ceroi.net

ZIMBABWE

BirdLife International species fact sheet on the African Fish-eagle and Bateleur. www.birdlife.org/datazone

Handbook of the Birds of the World, volume 2.

Zimbabwe Bird, Wikipedia

Image credits

Bloomsbury Publishers would like to thank the following for providing photographs and for permission to produce copyright material. While every effort has been made to trace and acknowledge all copyright holders, we would like to apologise for any errors or omissions and invite readers to inform us so that corrections can be made in any future editions of the book.

Photographs
Key t=top; l=left; r=right; tl=top left; tcl=top centre left; tc=top centre; tcr=top centre right; tr=top right; cl=centre left; c=centre; cr=centre right; b=bottom; bl=bottom left; bcl=bottom centre left; bc=bottom centre; bcr=bottom centre right; br=bottom right

G = Getty; FLPA = Frank Lane Picture Agency; SH = Shutterstock; NPL = Nature Picture Library; Wiki = Wikimedia

Cover front and back background Mikhail Pogosov/SH, Derek Middleton/FLPA, Jurgen & Christine Sohns/FLPA, Miguel Azevedo e Castro/SH, CreativeNature.nl/SH, 60293089/SH, H. Tuller/SH, Pablo Hidalgo/SH, holbox/SH, Joachim Hiltmann/Imagebroker/FLPA, Imagebroker, Stefan Huwiler/Imagebroker/FLPA, David Tipling/FLPA, Vishnevskiy Vasily/SH, Clinton Moffat/SH, Andy Sands/NPL, Cornelius Paas/Imagebroker/FLPA, Wolfgang Kruck/SH; **title page** Morphart Creation/SH; **6 bl** Mike Danzenbaker, bc Martin Hale/FLPA, **br** Forestry and National Parks Department, Anthony Jeremiah Grenada; **7 tl** Dubi Shapiro, **tc** Kim Taylor/NPL, **tr** Rod Williams/NPL, **cl** Daniel Heuclin/NPL, **c** Patricio Robles Gil/NPL, John Cox, **bl** Otto Plantema, **bc** Frank W Lane/FLPA, **br** David Tipling/FLPA; **9** Francis Bossé/SH; **10 tl** David Tipling/FLPA, **bl** Imagebroker, Stefan Huwiler/FLPA; **11** Staffan-Widstrand/ NPL; **12 tr** Globe Turner/SH, **tl** Gerard Soury/Biosphoto/FLPA, **cr** Norman Dijkstra/Angola Post, br John Hornbuckle; **13 tr** megastocker/SH, **tl** Michael Gore/FLPA, **br** Chris Gibbins/Anguilla Post; **14 tr** smilestudio/SH, **tl** Mogens Trolle/SH; **15 tl** Tim Laman/NPL, **tr** Stacy Funderburke/SH, Chris Gibbins/Antigua & Barbuda Post; **16 tr** Paul Stringer/SH, **tl** James Lowen/FLPA, **br** Luis César Tejo/SH; **17 tl** Hermann Brehm/NPL, **cr** Joab Souza/SH, **bc** ImageBroker/FLPA, **br** Chris Gibbins/Argentina Post; **18 tr** Globe Turner/SH, **tl** Jurgen & Christine Sohns/FLPA, **bl** Tom and Pam Gardner/FLPA; **19 tcr** BMCL/SH, **tr** Katie Green, National Museum of Australia, **bl** Phil Hill/SH; **20 tl** Jurgen & Christine Sohns/FLPA, **tr** Jurgen & Christine Sohns/FLPA, **bc** Yva Momatiuk & John Eastcott/Minden Pictures/FLPA; **21 tl** Janelle Lugge/SH, **tc** Cloudia Spinner/SH, **tr** Martin Willis/Minden Pictures/FLPA, **c** Houshmand Rabbani/SH, **bc** Susan Flashman/SH; **22 tr** Globe Turner/SH, **tl** Simon Litten/FLPA, **bl** Reinhard Hölzl/Imagebroker/FLPA, **br** Hans Dieter Brandl/FLPA; **23 tl** Cyril Ruoso/Minden Pictures/FLPA, **cr** Chris Gibbins/Austrian Post, **br** Rozhkovs/SH; **24 tr** Smilestudio/SH, **tl** Imagebroker/FLPA; **25 tl** Edward Myles/FLPA, **cl** Gerry Ellis/Minden Pictures/FLPA, **bc** Nigel Redman/Bahamas Post, **br** Bahamas National Trust Logo; **26 tr** Globe Turner/SH, **tl** Hanne & Jens Eriksen/NPL, **br** Chris Gibbins/Bahrain Post; **27 tr** Globe Turner/SH, **tl** Hanne & Jen Eriksen/NPL, **bl** Chris Gibbins/Bangladesh Post, **bc** AFP/Getty Images, **br** Wikimedia Doyel Chatwar; **28 tr** Smile studio/SH, **tl** Imagebroker, Horst Jegen/Imagebroker/FLPA; **29 tl** Philippe Clement/NPL, **tr** Michel Gunther/Biosphoto/FLPA, **bcr** Chris Gibbins/Belarus Postal Service, **br** Belarus Postal Service; **30 tr** Globe Turner/SH, **tl** Bugtiger/SH, **c** Dietmar Nill/NPL, **br** AFP/Getty Images; **31 tr** Atlaspix/SH, **tl** Eduardo Rivero/SH, Chris Gibbins/Belize Post; **32** Jurgen & Christine Sohns/FLPA; **33 tr** megastocker/SH, **c** Mike Danzenbaker, **br** Jeremy Madeiros, Department of Conservation Services; **34 tr** Jeremy Madeiros, Department of Conservation Services; **35 tc** Jeremy Madeiros, Department of Conservation Services, **bl** Mandy Shailer, Department of Conservation Services Government of Bermuda, br Chris Gibbins/Source: reproduced by permission of the Bermuda Philatelic Bureau; **36 tr** megastocker/SH, **tl** Richard Seitre/NPL; **37 tl** Jim Hallet/NPL, **tr** Alan Williams/NPL, **bl** Chris Gibbins/Bhutan Post, **br** Pedro Ugarte AFP/Getty Images; **38 tr** Artgraphixel.com/SH, **tl** Wikipedia Condor des Andes, **bl** Pete Oxford/NPL, **br** Jose B. Ruiz/NPL; **39 tr** Chris Gibbins Bolivian Postal Service, **c** James Brunker LatinContent/Getty Images, **bl** Brian Bahr/Getty Images; **40 tr** David Tipling/FLPA, **bl** Richard Du Toit/NPL; **41 tl** ImageBroker/Imagebroker/FLPA, **tr** Globe Turner/SH, **cr** Richard Du Toit/NPL; **42 tr** Malcolm Schuyl/FLPA, **b** Richard Du Toit/Minden Pictures/FLPA; **43 tr** Globe Turner/SH, **t** elnavegante/SH, Source: Brazilian Post and Telegraph Company; **44 tr** Atlaspix/SH, **tl** Rolf Nussbaumer/NPL, **br** Chris Gibbins/British Virgin Islands Post; **45 tr** Globe Turner/SH, **tl** Roland Seitre/NPL, **bl** bumihills/SH; **46 tr** Atlaspix/SH, **tl** Giant Ibis Martin Hale/FLPA; **47** Chris Gibbins/Cambodia Post; **48 tr** Atlaspix/SH, **tl** Jurgen & Christine Sohns/FLPA, **bl** Stuart Mailer, **bc** Philip Friskorn/Minden Pictures/FLPA; **49** Nigel Redman/Cayman Islands Post, Nigel Redman/Cayman Islands Post; **50 tr** Globe Turner/SH, **tl** David Tipling/FLPA, **bl** David Tipling/FLPA; **51 tr** Chris Gibbins/Chile Post, Condor online magazine; **52 tr** Paul Stringer/SH, **tl** Dickie Duckett/FLPA; **53** Konrad Wothe/Minden Pictures/FLPA; **54 tr** Nigel Redman/China Post, **cl** Keren Su/Getty, **bl** Lefteris Papaulakis/SH, **br** TAO Images Limited/Getty; **55 tr** Paul Stringer/SH, **tl** Murray Cooper/NPL; **56 tr** Atlaspix/SH, **cr** Eitan Abramovich AFP/Getty Images, **bl** Jean-Francois Deroubaix/ Gamma-Rapho via Getty Images; **57 tr** Guillermo Legaria/Stringer AFP/Getty Images, **bl** Veronique Durruty/ Gamma-Rapho via Getty Images, **br** Chris Gibbins/Colombia Post; **58 tr** Artgraphixel.com/SH, **tl** Franck & Christin/Biosphoto/FLPA, **br** Chris Gibbins/Costa Rica Post; **59 tr** Paul Stringer/SH, **tl** Nigel Redman, **br** Nigel Redman/Cuba Post; **60 tr** Fotogroove/SH, **tl** Paul Sawer/FLPA, **bc** Wim Reyns/Minden Pictures/FLPA; **61 tl** Mike Lane/FLPA, **cl** Malcolm Schuyl/FLPA, **br** British Library/Robana/British Library/Robana via Getty Images; **62 tr** Atlaspix/SH 00052495 Frank W Lane/FLPA, **tl** Stephen Dalton/NPL; 63 Tr Paul R. Reillo, Ph.D. **cl** Roland Seitre/NPL, Nigel Redman/Dominica Post, **bl** Roland Seitre/NPL; **64 tr** Globe Turner/SH, Francesco Tomasinelli, **bc** Chris Gibbins/Dominican Republic Post, **br** Neil Bowman/FLPA; **65 tr** Atlaspix/SH, **tl** Tui De Roy/Minden Pictures/FLPA; **66 tr** Alfredo Maiquez/Getty, **cl** Jim Clare/NPL, **bc** LatinContent/STR/LatinContent/Getty Images, **br** Nigel Redman/Ecuador Post; **67 tr** Atlaspix/SH, **tl** Michael Gore/FLPA, Maru Panameno Torogoz, S.A. de C.V., Nigel Redman/El Salvador Post; **68 tr** GlobeTurner/SH, **tl** Tony Hamblin/FLPA, **69 tr** Bill Baston/FLPA, **tr** Konrad Wothe/Minden Pictures/FLPA, **bcr** Bloomberg via Getty Images, **br** Chris Gibbins/Austrian Post; **70 tr** wavebreakmedia/SH, **tl** Bernard Castelein/NPL, **bc** Faroe Islands Postal Service stamps.fo, **br** Mike Powles/FLPA; **71 tr** Fotogroove/SH, **tl** Danny Green/NPL; **72 tr** Dieter Hopf/Imagebroker/FLPA, **cr** Paul Hobson/FLPA, **bl** ImageBroker/Imagebroker/FLPA; **73** Nigel Redman/© Itella Posti, Finland, **tcr** Image: Mint of Finland, **bl** stockbyte/Getty; **74 tr** Anatoly Tiplyashin/SH, **tl** Lynn M. Stone/NPL, **bl** Miguel Azevedo e Castro/SH, **bc** De Meester / ARCO/NPL; **75 tc** Dean Mouhtaropoulos/Getty Images, **tr** Lefteris Papaulakis/SH, **trr** Olemac/SH, **bl** le coq sportif, **br** Philippe Huguen/AFP/ Getty Images, **76 tr** megastocker/SH, **tl** Martin Woike/Minden Pictures/FLPA, Chris Gibbins/Gibraltar Post, **br** Roland Seitre/NPL; **77 tr** Atlaspix/SH, **tl** Forestry and National Parks Department, Anthony Jeremiah Grenada, Chris Gibbins/Grenada Post; **78 tl** Atlaspix/SH, **l** Imagebroker, Stefan Huwiler/Imagebroker/FLPA, **br** Imagebroker, Stefan Huwiler/Imagebroker/FLPA; **79** Patricio Robles Gil/NPL; **80 tl** Nigel Redman/Guatemala Post, **tr** Bocman1973/SH, **cl** Banco de Guatemala, **bc** Roland Seitre/NPL; **81 tr** Morphart Creation/SH, **bl** Bloomberg via Getty Images; **82 tr** Atlaspix/SH, **tl** Pete Oxford/NPL, **bl** SA Team/Minden Pictures/FLPA, **bc** Flip De Nooyer/FN/Minden/FLPA; **83 tr** Konrad Wothe/Minden Pictures/FLPA, **cl** Roger Tidman/FLPA, **cr** Chris Gibbins/Guyana Post, **br** Pete Oxford/NPL; **84 tr** Paul Stringer/SH, **tl** Gregory Guida/Biosphoto/FLPA, **bc** Chris Gibbins/Haiti Post, **br** Gregory Guida/Biosphoto/FLPA; **85 tr** pdesign/SH, **tl** Roland Seitre/NPL, **br** Frans Lanting/FLPA, **bc** Rod Williams/NPL; **86 tr** Dave Rock/SH, **bl** LoraKim Joyner; **87** LoraKim Joyner, LoraKim Joyner, LoraKim Joyner, Chris Gibbins/Honduras Post; **88 tr** Globe Turner/SH, **tl** Roger Tidman/FLPA, **bl** Ignacio Yufera/FLPA; **89 tl** Ignacio Yufera/FLPA, **cr** Nigel Redman/Hungary post, **br** MME logo www.mme.hu; **90 tr** Peter Wemmert/SH, **tl** Wild Wonders of Europe / Bergman/NPL; **91 tl** Konrad Wothe/Minden Pictures/FLPA, **cl** David Kjaer/NPL, **cr** Iceland Post, **bc** Roland Seitre/NPL; **92 tr** M. Shcherbyna/SH, **tl** David Noton/NPL, **bl** John Watkins/FLPA, **bc** Ignacio Yufera/FLPA; **93 tr** Keren S/Getty, **cl** UIG via Getty Images, **br** SSPL via Getty Images; **94 tl** Michael Gore/FLPA, **tr** Neil Bowman/FLPA, **bc** John Holmes/FLPA; **95 tc** David Hosking/FLPA, **bl** Neil Bowman/FLPA, **br** Imagebroker, Christian Hatter/Imagebroker/FLPA; **96 tr** Paul Stringer/SH, **tl** Dubi Shapiro, **br** Bas van Balen/BirdLife International; **97 tr** wiki government indonesia Garuda_Pancasila, **br** Adek Berry/Stringer/AFP/Getty, Chris Gibbins/Indonesia Post; **98 tr** AFP/Stringer/AFP/Getty Images, **cl** Bay Ismoyo/Stringer, **br** AFP/Getty

Images; **99 tr** Globe Turner/SH, **tl** Marcel van Kammen/Minden Pictures/FLPA, **bl** Dickie Duckett/FLPA; **100 tl** Mike Jones/FLPA, **tc** Roger Tidman/FLPA, **cr** Chris Gibbins/Israel Post, **bl** © Copyright 2012 The Bank of Israel, All Rights Reserved; **101 tr** Globe Turner/SH, **l** Neil Bowman/FLPA, **br** Rolf Nussbaumer/NPL; **102 tl** Rolf Nussbaumer/NPL, **tr** Nigel Redman/Jamaica Post, Money Museum, Bank of Jamaica; **103 tr** Globe Turner/SH, **tl** Shin Yoshino/Minden Pictures/FLPA, **br** Radu Razvan/SH; **104 tr** Globe Turner/SH, **tl** Oriol Alamany/NPL, **bl** Waj/SH, **bc** Chris Gibbins/Jordan Post; **105 tr** Globe Turner/SH, **tl** Steve Knell/NPL, **bc** Latvian Ornithological Society, **br** Chris Gibbins/Latvia Post; **106 tr** M. Shcherbyna/SH, **tl** Bernd Rohrschneider/FLPA; **107 tl** Markus Varesvuo/NPL, **tr** Jurgen & Christine Sohns/FLPA, Nigel Redman/Liberia Post; **108 tr** Paul Stringer/SH, **tl** Wild Wonders of Europe / Hamblin/NPL; **109 tc** Chris Gibbins/Lithuania Post, **tr** Lithuanian Ornithological Society, **cr** Lithuanian Ornithological Society; **110 tr** Globe Turner/SH, tl Markus Varesvuo/NPL, **br** Chris Gibbins/Source: Post Luxembourg; **111 tr** Globe Turner/SH, tl Simon_g/NPL, **bc** Grant Clow/NPL, **br** Sharon Heald/NPL; **112 tr** Tony Heald/NPL, **cl** Giovanni De Caro/SH, **br** © Biosphoto, **br** Bruno Cavignaux/Biosphoto/FLPA; **113 tr** Megastocker/SH, **tl** Hanne & Jens Eriksen/NPL, **bl** Bill Baston/FLPA; **114 tl** Michael Gore/FLPA, **tc** Richard Brooks/FLPA, Nigel Redman/Source: MaltaPost p.l.c; **115 tr** Globe Turner/SH, **tl** Natural History Museum (WAC)/NPL, **bc** Roger Tidman/FLPA; **116 tr** Gerard Lacz/FLPA, **cr** Cornelius Paas/Imagebroker/FLPA, **bl** Nigel Redman/Mauritius Post; **117 tr** Artgraphixel.com/SH, **tl** John Cancalosi/NPL, **bl** Tom Vezo/NPL; **118 tl** Hermann Brehm/NPL, **tr** Rolf Nussbaumer/NPL, **br** Danita Delimont/Getty; **119 tr** megastocker/SH, **tl** Kim Taylor/NPL, **bc** Xi Zhinong/NPL, **br** Eric Dragesco/NPL; **120 tr** Chris Gibbins/Mongolia Post, **cr** David Tipling/FLPA; **121 tr** megastocker/SH, **tl** Rod Williams/NPL, **bc** Roland Seitre/NPL, **br** Jurgen & Christine Sohns/FLPA; **122 tr** Stringer/AFP/Getty, **br** Chris Gibbins/Montserrat Post; **123 tr** Atlaspix/SH, **tl** Christian Heinrich/Imagebroker/FLPA; **124 tr** Martin B Withers/FLPA, **cr** Richard Du Toit/NPL, **bl** Namibia Rugby Union, **br** Atlaspix/SH; **125 tr** Atlaspix/SH, **cl** Patricio Robles Gil/NPL, **br** Patricio Robles Gil/NPL; **126 tl** Patricio Robles Gil/NPL, **br** Nigel Redman/Nepal Post; **127 tr** SmileStudio/SH, **tl** Daniel Heuclin/NPL; **128 tl** A© Biosphoto , Nicolas-Alain Pet/Biosphoto/FLPA, **br** Nigel Redman/New Caledonian Post; **129 tr** Globe Turner/SH, **tl** Terry Whittaker/FLPA; **130 tr** Mark Jones/Minden Pictures/FLPA, **c** Roger Tidman/FLPA, **bl** Tui De Roy/Minden Pictures/FLPA; **131 tr** Lasting Images/Getty, **c** CreativeNature.nl/SH, **bll** Asaf Eliason/SH, **bl** IgorGolovniov/SH, **br** NPL, **brr** Bank of New Zealand; **132 tr** Atlaspix/SH, **tl** Michael Gore/FLPA, Nigel Redman/Nicaragua Post; **133 tr** amorfati.art/SH, **tl** Mike Wilkes/NPL, **bl** Mike Wilkes/NPL; **134 tl** Neil Lucas/NPL, Chris Gibbins/Nigeria Post; **135 tr** Fotogroove/SH, **tl** Andy Rouse/NPL, Chris Gibbins/Norway Post; **136 tr** Iakov Filimonov/SH, **tl** Tom Reichner/SH, **br** Michael Szönyi/Imagebroker/FLPA, **bc** Chris Gibbins/Pakistan Post; **137 tr** Darren Whittingham/SH, **tl** Margaret Sloan, **br** Margaret Sloan; **138 tr** Tim Laman/Getty, **cr** Chris Gibbins/Palau Post, **br** Palau Conservation Society; **139 tr** Darren Whittingham/SH, **tl** David Hosking/FLPA, **br** David Hosking/FLPA; **140 tl, cl,** NGArchitects, **br** Atlaspix/SH; **141 tr** Vanatchanan/SH, **tl** Tui De Roy/Minden Pictures/FLPA; **142 tr, br** Images by David N. Kitler, resulting from his trip to Panama during the first-ever Artists for Conservation Flag Expedition; **143 tr** Globe Turner/SH, **tl** Tim Laman/NPL; **144 tl** David Tipling/FLPA, **b** Phil Savoie/NPL; **145 tr** Air Niugini, **bl** Charles Lagus/Getty, Nigel Redman/Papua New Guinea; **146 tr** Paul Stringer/SH, **tl** Nick Gordon/NPL, Fauna Paraguay, www.faunaparaguay.com; **147 tr** Globe Turner/SH, **tl** Kevin Schafer/NPL, **bl** Pete Oxford/NPL; **148 tl** Pete Oxford/Minden Pictures/FLPA, **tr** Mark Newman/FLPA, **br** Hermann Brehm/NPL, Chris Gibbins/Peru Post; **149 tr** Smilestudio/SH, cl Patricio Robles Gil/NPL; **150 tr** Noel Celis/AFP/Getty Images, **bl** Romeo Gacad/AFP/Getty Images, **br** Romeo Gacad/AFP/Getty Images; **151 tr** Atlaspix/SH, **cl** David K. Disher, **br** Roland Seitre/NPL; **152** Raul Quinones; **153 tr** megastocker/SH, **tl, br, bc** John Cox; **154 tr** Wikimedia Gustav Mützel, public domain, **br** Chris Gibbins/Samoa Post; **155 tr** PromesaArtStudio/SH, **tl** Mark Caunt/SH; **156 cl** Mark Caunt/SH, **cr** Matt Gibson/SH, **bl** Philip Ellard/SH, **br** Ewan Chesser/SH; **157 tr** Globe Turner/SH, **tl** David Pike/NPL; **158 t, bl** Seychelles Islands Foundation (SIF), Chris Gibbins/Seychelles Post; **159 tr** SmileStudio/SH, **cl** Chien Lee/Minden Pictures/FLPA, **bl** Roland Seitre/NPL, **bc** isarescheewin/SH; **160 tr** cowboy54/SH, **br** Nature Society (Singapore) logo; **161 tr** Route66/SH, **tl** Malcolm Schuyl/FLPA, **br** Malcolm Schuyl/FLPA; **162 tl** Edward Myles/FLPA, **cr** Hulton Archive/Getty Images, **bl** Overberg Crane Group; **163 tr** amorfati.art/SH, **tl** Kevin Schafer/NPL, **bc** Nigel Redman/Sri Lanka Post, **br** Simon Hosking/FLPA; **164 tr** Laschon Robert Paul/SH, **tl, br** Otto Plantema; **165 tr** St Helena National Trust, **cl** St Helena National Trust/Getty; **br** Chris Gibbins/St Helena Post; **166 tr** Kheng Guan Toh/SH, **tl** jo Crebbin/SH, **bc** Laschon Robert Paul/SH, **br** Tom Vezo/NPL; **167 tr** Globe Turner/SH, **cl** Rod Williams/NPL, **br** Krystyna Szulecka/FLPA; **168 tr** Laschon Robert Paul/SH, **bl** Chris Gibbins/St Lucia Post; **169 tr** Globe Turner/SH, **tl** tentpole/istock/Getty; **170 tr** Robin Budden/NPL, Nigel Redman/St Vincent Post; **171 tr** dicogm/SH, **tl** Mike Wilkes/NPL, **bl** Edwin Giesbers/NPL, **bc** Chris & Tilde Stuart/FLPA; **172 tl** Barry Bland/NPL, **tr** Mary McDonald/NPL, **br** Ashraf Shazly/ Stringer/AFP/Getty Images; **173 tr** jurie/SH, **tl** Ann & Steve Toon/NPL, **bc** Chris & Tilde Stuart/FLPA; **174 tl** Philip Perry/FLPA, **cl** Neftali/SH, **bl** Chris Gibbins/Swaziland Post; **175 tr** Globe Turner/SH, **tl** Andy Sands/NPL, **bc** Chris Gibbins/ © Sweden Post Stamps, **br** Paul Miguel/FLPA; **176** George McCarthy/NPL; **177 tr** amorfati.art/SH, **tl** del.Monaco /SH, **bc** Taiwan International Birding Association (TBA) Formosan Magpie, **br** Chi-An/SH, **178 tr** Globe Turner/SH, **cl** gopause/SH; **179 tr** Roland Seitre/NPL, **cr** Roland Seitre/NPL, **br** Wikimedia: Didier Descouens, **bl** Chris Gibbins/Thailand Post; **180 tr** Tim Laman/NPL, **bl** ImageBroker / Imagebroker /FLPA; **181 tl** Murray Cooper/NPL, **tr** Globe Turner/SH, **bl** Comstock Images / Getty Images; **182 tr** megastocker/SH, **tl** Suzi Eszterhas/NPL, **bl** Digfoto/Imagebroker/FLPA; **183 tl** Imagebroker, Siegfried Kuttig/Imagebroker/FLPA, tc Michel and Christine Denis-Huot/Biosphoto/FLPA, **cr** William West/AFP/Getty Images, **bl** Chris Gibbins/Uganda Post; **184 tr** charnsitr/SH, **tl** Rob kemp/SH, **bl** Derek Middleton/FLPA, **bc** Robin Chittenden/NPL; **185** T.J. Rich/NPL; **186 tr** UniversalImagesGroup/Getty Images, **cl** The National Archives/SSPL via Getty Images; **187 tl** UniversalImagesGroup/Getty Images, **br** Hulton Archive/Getty Images; **188 tr** megastocker/SH, **tl** Nigel Bean/NPL, **br** Rolf Nussbaumer/Imagebroker/FLPA; **189 tl, tr** Rolf Nussbaumer/NPL, Wikimedia: Escondites/Public Domain US Government; **190 tr** Stefanina Hill/SH, **tl** David Tipling/NPL; **191 tr** Michael Quinton/Minden Pictures/FLPA, **tl** Steve Gettle/Minden Pictures/FLPA; **192 tr** Werner Forman/UIG via Getty Images, **cr** Maria Toutoudaki Getty Images, **bl** Olga Popova/SH, **bcl** stamp traveler1116/Getty, **bc** Siede Preis/Getty; **193 tl** Peter Llewellyn/FLPA, **tc** Konrad Wothe/Minden Pictures/FLPA, **tr** Donald M. Jones/Minden Pictures/FLPA, **bc** Michael Quinton/Minden Pictures/FLPA; **194 tl** Scott Leslie/Minden Pictures/FLPA, **tc** S & D & K Maslowski/FLPA, **tr** S & D & K Maslowski/FLPA, **bc** Kevin Schafer/Minden Pictures/FLPA; **195 tl** Donald M. Jones/Minden Pictures/FLPA, **tc** Bill Coster/FLPA, **tr** S & D & K Maslowski/FLPA, **br** Hugh Clark/FLPA; **196 tl** Daphne Kinzler/FLPA, **tc** Rolf Nussbaumer / Imagebroker / FLPA, **tr** Jim Brandenburg/Minden Pictures/FLPA, **bl** S & D & K Maslowski/FLPA; **197 tl** Scott Leslie/Minden Pictures/FLPA, **tc** Tom Vezo/Minden Pictures/FLPA, **tr** Cyril Ruoso/Minden Pictures/FLPA, **bc** S & D & K Maslowski/FLPA; **198 tl** Donald M. Jones/Minden Pictures/FLPA, **tc** Mark Raycroft/Minden Pictures/FLPA, **tr** S & D & K Maslowski/FLPA; **199 tl** Malcolm Schuyl/FLPA, **tc** Tom Vezo/Minden Pictures/FLPA, **tr** Bill Coster/FLPA, **br** Dennis Lorenz/Minden Pictures/FLPA; **200–201** Nigel Redman/ State Birds & Flowers © 1982 United States Postal Service. All Rights Reserved. Used with Permission; **202 tr** Paul Stringer/SH, tl Murray Cooper/Minden Pictures/FLPA, **br** Claus Meyer/Minden Pictures/FLPA; **203 tl** wikimedia Bjørn Christian Tørrissen, **tr** wikimedia Lswarte, **br** Chris Gibbins/Venezuela Post; **204 tr** Globe Turner/SH, **cl** Hanne & Jens Eriksen/NPL, **br** Hanne & Jens Eriksen/NPL; **205 tl** javarman/SH, **tr** Fabio Pupin/FLPA, **bl** Atlaspix/SH, **br** Yossi Eshbol/FLPA; **206 tr** megastocker/SH, **tl** Frans Lanting/FLPA, **br** Christian Heinrich/Imagebroker/FLPA; **207 tl** Tim Fitzharris/Minden Pictures/FLPA, **cr** Nigel Redman/Zambia Post; **bl** Alexander Joe/AFP/Getty Images, **br** Comstock Images/Getty Images; **208 tr** PromesaArtStudio/SH, **tl** Sharon Heald/NPL, **br** John Downer/NPL; **209 tl** Richard Du Toit/NPL, **tr** Tony Heald/NPL, **cr** Chris Gibbins/Zimbabwe Post, **br** PromesaArtStudio/SH; **210 tl** Atlaspix/SH, **tc** Atlaspix/SH, **tr** Paul Reeves Photography/SH, **211 tl** Arto Hakola/SH, **tc** Brian Lasenby/SH, **tr** Paul Reeves Photography/SH, **ctl** Donya Nedomam/SH, **ctc** Tom Reichner/SH, **ctr** /Imagebroker/FLPA, **bl** Ronnie Howard/SH, **bc** Pim Leijen/SH, **br** Mark Caunt/SH, **bbl** Elliotte Rusty Harold/SH, **bbc** Stanislav Duben/SH, **bbr** Stanislav Duben/SH; **212 tr** Iakov Filimonov/SH, **cr** Jonathan Noden-Wilkinson/SH, **bl** F1online/F1online/FLPA, **br** Rozhkovs/SH; **213 tl** Laschon Robert Paul/SH, **tc** Rozhkovs/SH, **tr** Erni/SH, **cr** yui/SH, **cr** Marcus Siebert/Imagebroker/FLPA; **214 tl** smishonja/SH, **tr** svic/SH **bl** Atlaspix/SH, **bc** Dvorko Sergey/SH, **br** David Hosking/FLPA; 215 tl Anton_Ivanov/SH, **tr** Laschon Robert Paul/SH, **cr** Vitaly Ilyasov/SH, **br** Eduardo Rivero/SH

Index

Q

R

S

T

U

V

W

Y

Z